Nucleic Acid Biochemistry

Nucleic Acid Biochemistry

Robert Wilson

Oxford Book Company
Jaipur ▪ New Delhi

Published By

Oxford Book Company

Regd. Off. "Bony Residency", Gate No. 2
Opp. Tilak Public School, Vishweswariya Nagar,
Gopalpura Road, Jaipur - 302018, India
Ph: 0141 - 2761280, Telefax: 2761381
email - oxfordbookjpr@gmail.com
website - oxfordbookcompany.com

Delhi Office: 102, 1st Floor, Satyam House,
4327/3, Ansari Road, Darya Ganj,
New Delhi - 110002, India
Ph: 011-45652440

Distributed By

KEYA INTERNATIONAL PRESS LLP
E-mail: keyainternationalpressllp@gmail.com

First published in India, 2026

ISBN : 978-93-5524-226-6

Typeset by Shivangi Computers, Jaipur

Printed and bound in India by Thomson Press Ltd.,
New Delhi

Contents

Preface

Life, in all its grandeur and complexity, unfolds through a coded script etched within every living entity. This script is written not with ink but in molecules—nucleic acids—that govern the blueprints of existence. These molecular scribes carry the complete instructions for crafting life, orchestrating both the replication of organisms and the production of biomolecules essential to survival.

At the heart of this system are two remarkable molecules: deoxyribonucleic acid (DNA) and ribonucleic acid (RNA). While multicellular organisms rely predominantly on DNA to store and transmit genetic information, simpler life forms like viruses often turn to RNA to fulfill this role. Structurally distinct yet functionally profound, RNA typically exists as a single-threaded strand, while DNA elegantly twists into a double-helix, each strand perfectly matched to its partner through precise base pairing.

Despite their molecular elegance, nucleic acids are built from surprisingly few ingredients. Just four nucleotide building blocks form the foundation of each strand—adenine (A), guanine (G), cytosine (C), and thymine (T) for DNA, with uracil (U) replacing thymine in RNA. Yet, these minimal units create an expansive and intricate biological language. Within this language lie genes—compact sequences of nucleotides—each representing a distinct message that can be read and translated into specific cellular components.

Astonishingly, this quartet of bases constructs the full encyclopaedia of life. Consider the human genome: a sprawling sequence of over 3.2 billion nucleotide pairs. Even simpler organisms like Haemophilus influenzae manage all their life processes using just 1.8 million of these base pairs. Much like how 26 letters of the English alphabet can craft a near-infinite array of words and meanings, these four biological letters assemble the myriad instructions for life's diversity.

But it's not merely the presence of these sequences that matters—it is their precise arrangement. A slight alteration in the sequence can yield dramatically different outcomes. A gene beginning with "CCTTGGAA" might code for a completely different protein than one beginning "AATTGGAA," despite their superficial similarity. Accuracy here is vital; even a single error in sequencing can compromise an organism's ability to create vital biological structures.

Nucleic acids don't merely store data—they execute it. They act as the biological operating system, overseeing every aspect of cellular behavior. They carry the code that enables life to reproduce, evolve, and adapt. Without them, life as we know it would simply collapse. From encoding enzymes and hormones to enabling species inheritance, these molecules are the foundational architects of all living systems.

Beyond the primary sequence—a linear arrangement of nucleotides—these molecules also exhibit complex three-dimensional configurations, known as secondary and tertiary structures. These forms are not ornamental; they are functional blueprints that determine how nucleic acids interact within the cell and with other biomolecules.

Each nucleotide is a trifecta of structure and function: a sugar, a phosphate group, and a nitrogenous base. Whether ribose in RNA or deoxyribose in DNA, the sugar forms the backbone of the chain, holding the molecule together with chemical precision. Attached to each sugar is a base—either purine or pyrimidine—that acts as the language unit in this molecular script.

This book, Nucleic Acid Biochemistry, is crafted to offer a deeply engaging yet accessible exploration of this fundamental biological subject. It is an indispensable guide for students and professionals across medical sciences, microbiology, biotechnology, and pharmaceuticals. Designed with both clarity and comprehensiveness, this book serves as both a primer and a reference, capturing the ever-expanding horizon of nucleic acid research while firmly grounding the reader in essential principles that define the molecular essence of life.

— ***Robert Wilson***

Chapter

1

Simple Molecules

We have seen that an atom of an element has a characteristic number of electrons, equal to its atomic number Z. These electrons are arranged in shells about the nucleus, some being close to the nucleus, others far away. It is the electrons in the outermost shell, furthest from the nucleus, and therefore least tightly held, that are involved in chemical bonding. These are the valence electrons.

The electrons close to the nucleus, called inner or core electrons, are much too tightly held to be lost to or shared with another atom. As we have seen, the number of valence electrons that an atom has is equal to the last digit of its group number. Thus Na, in group 1, has one valence electron; Al in group 13 has three valence electrons; Br in group 17 has seven valence electrons; and Fe in group 8 has eight valence electrons. The Lewis symbol for an element shows the valence electrons as dots arranged about the symbol for the element, for Na, Al, and Br. Dots are placed as if the symbol of the element were surrounded by an invisible square.

Valence Electrons of Atoms

The first dot can be placed on any of the 4 sides, but the second, third, and fourth dots are then placed singly on the remaining 3 sides of the square; electrons are not placed in pairs (*i.e.*, 2 dots on the same side of the square) until 4 have been placed singly. Dots that singly occupy the side of a square are referred to as unpaired electrons.

G.N. Lewis developed this approach to representing the valence electrons of an element before the quantum view of the atom was complete. It is of interest from our perspective, though, to examine the correspondence between the Lewis symbol for an element and its valence electron configuration. In particular, we might expect the number of unpaired electrons in the Lewis symbol to correspond with the number of electrons that singly occupy valence orbitals of the atom. The Lewis symbols and valence orbital occupations of the atoms of period 2.

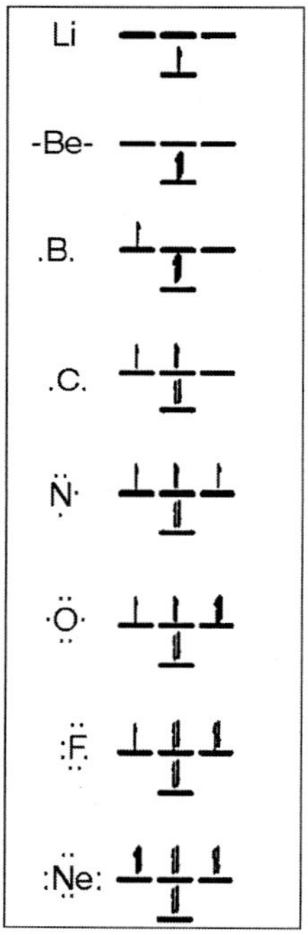

Figure: *Valence Electrons for Period2*

What we see is interesting and a bit puzzling. First, most of the atoms show the intuitively expected correspondence. Thus Li, N, O, F, and Ne exhibit the same number of unpaired electrons in the Lewis symbol and in the quantum orbital occupation diagram. Further, the numbers of unpaired electrons for these elements correspond with their usual valences; *i.e.*, with the number of bonds that they form.

However, for Be, B, and C in groups 2, 13, and 14, the numbers of unpaired electrons in the Lewis symbols do not correspond with the orbital occupation diagrams. Thus the Lewis symbol for carbon leads us to expect 4 unpaired electrons; the orbital occupation diagram predicts 2. The observed valence for carbon is (almost) invariably 4, in accord with the Lewis symbol. Similarly, the valences of Be and B are as expected from the Lewis symbols, but in conflict with the orbital occupation diagrams. Yet the success of the quantum theory of the atom has been so profound that we believe completely that the orbital occupation diagrams are correct.

How are we to explain the bonding properties of elements in groups 2, 13, and 14 in terms of the orbital occupation diagrams? The orbital diagrams can be brought into correspondence with the observed valences by a simple expedient: the downspin electron of the pair in the s orbital is placed in an empty p orbital, with its spin parallel to those of the other electrons. When this is done for Be, B, and C, the numbers of unpaired electrons correspond with the predictions of the Lewis symbol, and with the observed valences. Thus the elements of groups 2, 13, and 14 behave as if their orbital occupation.

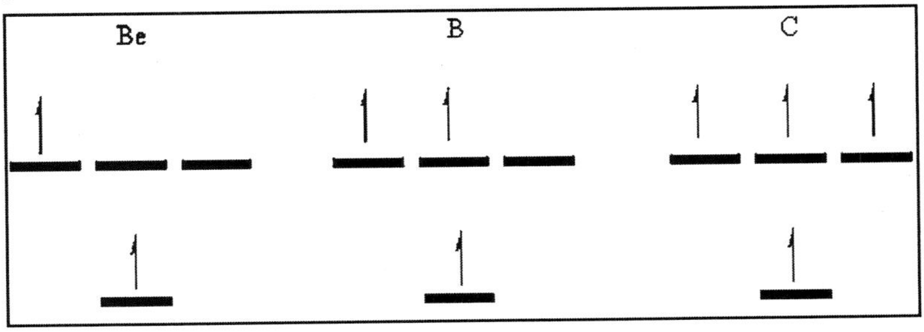

Figure: *Apparent Configurations for Be, B, and C*

The elements in group 18, called the Noble Gases, have 8 valence electrons, arranged in pairs on the 4 sides of the Lewis symbol square. The one exception is helium, He, which has 2 valence electrons. As their name indicates, these elements occur in nature as very unreactive gases; this implies that 8 valence electrons, called an octet, is in some sense a stable situation. As we saw that, 8 electrons is sufficient to fill the s and p orbitals of a shell.

When the atoms of the elements undergo chemical reactions, the frequent result is a filled set of valence s and p orbitals for each atom involved — *i.e.*, each atom acquires the octet of valence electrons possessed by the noble gases. The sodium atom usually loses its single valence electron upon reaction, forming Na^+. The sodium ion has the same number of electrons as the noble gas, Ne. The Al atom must lose 3 valence electrons to achieve a noble gas configuration, so it forms the Al^{3+} ion. In contrast, Br may much more readily gain a single electron to achieve the same number of electrons as Kr than lose 7 electrons. It therefore is found frequently in compounds as the

Br- ion. Elements with from 3 to 7 valence electrons often achieve an octet of electrons—a full set of valence s and p orbitals—by sharing, rather than gaining or losing, electrons. For example, the Br atom can achieve an octet if it can gain 1 electron by sharing. A second Br atom can supply this electron, and achieve the octet for itself in the process. Shared electrons contribute to the octets of both of the two sharing atoms.

The shared pair of electrons, located between the bromine atoms, is circled. Note that the single unpaired electron of one Br atom is paired with the single unpaired electron of the second Br atom. This pair is placed between them to indicate the sharing. The shared electron pair constitutes the bond between the two atoms, because it is attracted to both nuclei and therefore holds them together. The bond is called a covalent bond, because it involves cooperation between the 2 atoms so that each may achieve an octet. It is customary and convenient to represent electron pairs with lines rather than dots. Keep in mind that each line represents 2 electrons an electron pair.

Frequently, the number of covalent bonds that an atom forms is the same as the number of unpaired electrons in its Lewis symbol (we will see subsequently that many atoms can be represented by more than one Lewis symbol, giving them some flexibility in the number of covalent bonds formed). Therefore Br, and the other halogens, commonly form one covalent bond. Oxygen has two unpaired electrons and can achieve an octet by sharing both of these with another oxygen atom. The two oxygen atoms then share a total of four electrons, each thereby achieving the octet.

Two pairs of electrons shared between two atoms gives two bonds; the pair of bonds together is called a double bond, and is represented by a double line as in the structure for O_2. You should satisfy yourself that two nitrogen atoms can achieve an octet by sharing three pairs of electrons to form a triple bond. Carbon, with 4 unpaired electrons, forms 4 covalent bonds. Similarly, sulfur forms 2, phosphorus 3, silicon 4, hydrogen 1, and so on, based on the dot. Hydrogen is an exception to the octet rule, because it can accommodate only 2 valence electrons. In acquiring 2, it achieves the electron configuration of the noble gas, helium.

Lewis Structures for Molecules

The realization that when elements react to form compounds the atoms frequently achieve stable electron arrangements octets was first made by G.N. Lewis, a great American chemist. In tribute to the contributions made by Lewis to the theory of chemical bonding, molecular structures in which valence electrons are represented by dots or lines are called Lewis structures. The Lewis approach to structure, called Lewis theory, is our most-used theory of molecular structure. It is simple, yet sophisticated enough to provide an understanding of molecular properties such as shape, bond length, bond angle, and bond strength.

Further, it allows explanation of chemical reactivity and mechanism in terms of structure for many chemical reactions, and suggests a general approach to acids and bases that is used widely. In due course, we will address all of these matters. In this section, we develop a systematic approach for obtaining the Lewis structures of molecules. Some Lewis structures are easily obtained from the dot pictures of the constituent atoms.

Dot pictures for carbon and fluorine. Clearly carbon is able to form four bonds, and each fluorine is able to form one. Therefore, one C atom bonds to four F atoms. The Lewis structure for CF_4, tetrafluoromethane. Similarly, the Lewis structures for ammonia and water are simply obtained. In these molecules, each atom forms a number of covalent bonds equal to the number of unpaired electrons in its dot picture, and each atom has an octet (a duet in the case of hydrogen, which gives it the configuration of He, with 2 valence electrons) in the final structure.

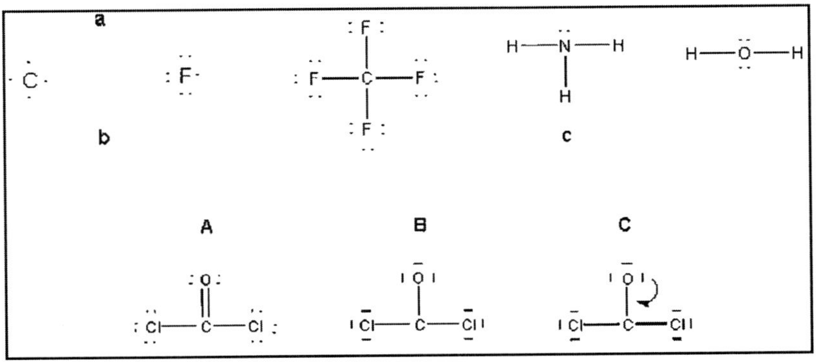

The Lewis structure is not always so readily obtained, however. For example, that for phosgene, $COCl_2$. Although inspection of this structure verifies that each atom has an octet and has formed the expected number of bonds, arriving at this structure from scratch requires some thought. Note the presence of a double bond between the carbon and oxygen atoms. How are we to know when double bonds and triple bonds, collectively called multiple bonds, are appropriate? To address this question, we now present a systematic and straightforward approach to the writing of Lewis structures.

- *Step:* Arrange the atoms in the correct layout (*i.e.*, bonding arrangement). The layout is experimentally determined, but can often be deduced from the formula for the molecule (or ion). For many molecules (but not all!), the first listed atom is in the center, and the others are attached to it. This is the case for phosgene.
- *Step:* Count the number of valence electrons in the molecule by summing the electrons contributed by each atom, based on their group numbers. If the molecule is negatively charged, add the corresponding number of electrons. Subtract the appropriate number of electrons if the molecule is positively charged. For phosgene,

Number of valence electrons = 6(O) + 4(C) + 2 × 7(Cl) = 24.

- *Step:* Develop a trial structure for the molecule by putting a single bond (1 electron pair, represented by a single line) between each pair of bonded atoms. Then complete the octets of as many atoms as possible by adding nonbonding electron pairs (often called lone pairs) to the atoms in turn until all of the available valence electrons (step 2) have been used. Situations will arise in which electrons remain to be added to the structure after the octets of all atoms are complete. In these situations, the remaining electrons are placed in pairs on the central atom.
- *Step:* Examine the trial structure. If all atoms have octets (or if the terminal atoms have octets while the central atom has more than 8 electrons), the trial structure is acceptable. If the trial structure shows atoms with fewer than eight electrons, it must be altered by using lone pairs to form multiple bonds until all atoms have octets.

The phosgene structure must clearly be amended because the carbon atom has only six electrons in the trial structure. A lone pair from the oxygen atom is moved in to form a second bond, preserving the octet on oxygen and at the same time establishing the octet for carbon. We emphasize that once the correct number of electrons is placed in the structure in step 3, no more may be added, and none may be removed; electrons can, however, be moved. (A lone pair of oxygen is preferred over one from chlorine for reasons that will be explained in a later section dealing with the concept of formal charge.) The result is an acceptable Lewis structure for phosgene. It is important to state at this point that an acceptable structure obtained by direct application of the procedure above is not necessarily the optimal (best) structure.

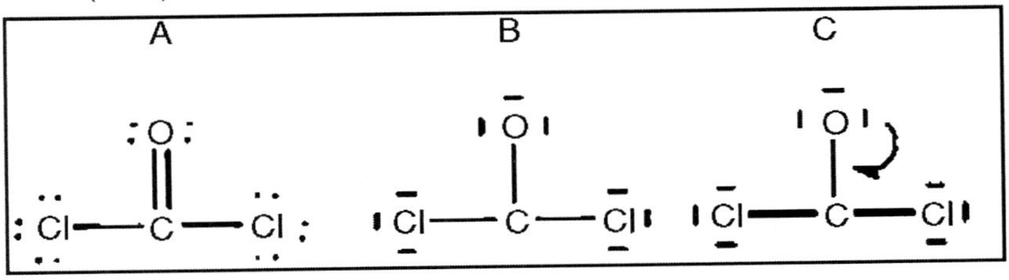

Exceeding the Octet

Although we have used the octet rule as the basis for development of a systematic approach to Lewis structures, only the atoms of period 2 of the periodic table obey it (almost) without exception. Any atom having valence shell principal quantum number ^{3} 3 may exceed the octet (violate the octet rule). When the octet of electrons is exceeded, it is usually at the central atom. An atom can exceed the octet only if it has atomic orbitals available for housing the additional electrons. As we have stated, eight electrons is sufficient to fill the s and p subshells of the valence shell (outermost main shell) of the atom. To accommodate more than eight electrons in the valence shell, an atom must have available orbitals in addition to the s and p. Atoms for which the valence shell has n ^{3} 3 have a valence d subshell comprised of five d-type atomic orbitals. When an atom exceeds the octet, it makes use of d orbitals to house the additional electrons.

The Se atom in SeF_4 uses a total of 5 valence orbitals (one s, three p, and one d) to accommodate the four bonding pairs and one lone pair of electrons that surround it. The sulfur atom in SF_6 uses 6 valence orbitals (one s, three p, and two d) to accommodate the six bond pairs. And the Te atom in TeF_8^{2-} uses 8 valence orbitals (one s, three p, and four of the five d) to house eight electron pairs! Note that we would not expect an atom with valence shell n ^{3} 3 to be able to accommodate more than 9 electron pairs in its valence shell because it has only this number of orbitals available in the s, p, and d subshells. Similarly, elements from period 2 can not house more than 4 electron pairs, because the n = 2 shell has no d subshell.

The atoms in period 2 are strictly limited by the octet rule. Many atoms can be represented by more than one Lewis symbol. This is very closely connected with the availability of d orbitals in the valence shell, and we will now examine this matter before moving on. The sulfur atom is a good vehicle for illustration, because it shows a substantial amount of flexibility in the number of covalent bonds that it is able to form. The usual Lewis symbol for S, and the orbital occupation diagram with which it corresponds. This Lewis symbol allows us to rationalize the many 2-valent compounds that are known for sulfur, among them H_2S, SCl_2, and $(CH_3)_2S$. Formation of two covalent bonds is readily rationalized in terms of the two unpaired electrons in the Lewis symbol (and orbital occupation diagram). Thus in H_2S, each H atom pairs its single unpaired electron with one of the unpaired electrons of S, resulting in completed s and p subshells for S, and completed s subshells for H.

The other two-valent molecules are understood in similar ways. However, sulfur also forms a number of four-valent compounds (for example, SF_4); and six-valent compounds (for example, SF_6), which cannot be readily accounted for in terms of the simple Lewis symbol above. However, useful Lewis symbols for S can be obtained by unpairing electrons, one pair at a time. The resulting Lewis symbols then show four and six unpaired electrons. Using these expanded Lewis symbols, we readily account for the flexivalent nature of the sulfur atom, and in fact for similar flexibility for most atoms with valence shell n ^{3} 3. The following generalization will prove useful to us in rationalizing many molecular structures:

A p-block atom usually forms compounds with a number of covalent bonds lying in the range between the number of unpaired electrons in its simplest Lewis symbol, and the total number of valence electrons that it has. Further, other common numbers of covalent bonds lie between these extremes, separated by increments of 2.

Applying this to S, we would predict that it should exhibit a minimum of 2 covalent bonds and a maximum of 6, with compounds containing 4 covalent bonds being common. Applying the guideline to chlorine enables us to predict a minimum of 1, a maximum of 7, with 3 and 5 covalent bonds being common. In fact, the compounds HCl (minimum), ClF_7 (maximum), ClF_3, and ClF_5 are well known.

A Formal Charge Refinement—Optimizing the Lewis Structure

We can combine our formal charge criteria for acceptable Lewis structures with the ability of many (central) atoms to exceed the octet to obtain improved structures for molecules in which the central atom is from period $n > 3$. By way of illustration, consider the structure of sulfuric acid. This structure results from application of the systematic approach to development of Lewis structures presented earlier. Although the sulfur and oxygen atoms obey the octet rule, the formal charges are a bit extreme.

Specifically, S has 2+ formal charge, and the two terminal oxygen atoms not bonded to hydrogen have formal charges of 1-. By using lone pairs on these two oxygens to form double bonds to S, the formal charges of all three atoms can be reduced to zero. This causes the sulfur atom to exceed the octet, but this is acceptable for an atom from period 3. The resonance form called the minimum formal charge structure, results. It is this structure that is usually drawn for sulfuric acid.

The minimum formal charge structure for any molecule can be obtained in a straightforward manner. Starting with the structure generated using the systematic procedure, assign formal charges to all atoms. If the central atom is not from period 2, introduce double bonds between terminal atoms of negative formal charge and the central atom until the central atom formal charge is zero. One double bond removes one unit of positive formal charge from the central atom, so to achieve zero formal charge requires a number of double bonds equal to the positive formal charge in the initial structure. Initial and minimum formal charge structures for H_4SiO_4, H_3PO_4, and $HClO_4$ are shown.

Bond Angle Refinements

The VSEPR Theory allows us to easily predict the overall shape of a simple molecule based on the Lewis structure, and bond angles for those highly symmetrical molecules in which all central atom electron groups are shared with terminal atoms, and in which all terminal atoms are the same. With the addition of just a few more simple ideas, the theory can be extended to allow us to say something about bond angles in less symmetrical situations as well.

We begin with some experimental facts. The electron group distributions in methane, ammonia, and water are tetrahedral. However, whereas the bond angle in methane is the expected 109.5°, that in ammonia is 107°, and that in water is still smaller at 104°. Why? The nonbonding electron pair that occupies one of the four positions around the nitrogen atom in ammonia belongs entirely to the nitrogen atom; it is not shared with any other atom. It is therefore closer to nitrogen than are the 3 single bond pairs, and requires more room. The result is that the three bond electron pairs are repelled more by the lone pair than by each other, and are squeezed closer togther than in methane; the bond angle is reduced. The effect is even greater in water, where two of the four groups are nonbonding pairs. Similarly, it is found experimentally that the angle between the two C-F bonds in COF_2 (similar to phosgene, $COCl_2$) is 108°, noticeably less than the ideal value of 120°.

This is attributed to the double bond between C and O, which involves four electrons rather than two, and therefore takes up more space. Finally, the O-N-O angle in NO_2 is 132°, somewhat larger than the ideal value of 120°. As might be expected, the single electron group on the central N atom exerts less repulsion than would a nonbonding pair, and less even than a bond pair of electrons. As a consequence, the angle between bond pairs opens up.

Based on these observations, the following generalizations may be made:

- Repulsion between lone pairs and bond pairs falls in the following order: LP-LP > LP-BP > BP-BP (LP stands for lone pair, BP for bond pair).
- Repulsion between multiple bonds and single bonds falls in the following order: MB-MB > MB-BP> BP-BP.
- Repulsion between a bond pair and a single electron is less than that between two bond pairs: BP-BP> BP-SE (SE stands for single electron).

With these ideas in hand, we now address the strange structures for SF_4 and BrF_3 in Table 3-2. There are 5 electron groups around the central atom in both species, so the group distribution is trigonal bipyramidal. In both cases, lone pairs, which exert the largest repulsions on other electron pairs, occupy equatorial rather than axial sites because by so doing they minimize 90° interactions with other electron pairs.

A lone pair in an axial position would have three such interactions (with the equatorial pairs); in an equatorial position, it has only two 90° interactions (with the axial pairs). Lone electron pairs occupy equatorial sites in the trigonal bipyramid. Similar arguments apply for square planar XeF_4, in which the two lone pairs are as far apart as possible at 180°.

Example: Say as much as possible about the bond angles and lengths in the following species: N_2O, SO_4^{2-}, SO_3^{2-}, O_3 (ozone), ClO_2.

Charge Distribution in Molecules

The electronegativity of an atom was defined by Pauling as "the ability of the atom in a molecule to attract electrons to itself." Unlike the molar volume and ionization energy, the electronegativity is not a directly measurable physical property of an atom. Rather, it is defined in a somewhat vague and qualitative way. Several ways of assigning a numerical value to the electronegativity of an atom have been developed. The most commonly used scale of electronegativity is that of Pauling, based on the experimentally measured values of bond energies. We will not concern ourselves with the details of how the scale was established.

We will instead present the scale, make some general statements about the variation of electronegativity with position in the periodic table, and then proceed to use the electronegativity where appropriate. Suffice it to say at this point that the electronegativity of an atom, symbolized c, is roughly proportional to the first ionization energy of the atom. In other words, the more difficult it is to remove the outermost electron from an atom, the more likely it is that the atom will attract additional electrons to itself.

The Pauling electronegativities of the elements are presented.

The following trends can be extracted:

- χ increases left-to-right across a period. This is just what we expect given the parallel between χ and I_1. However, there are no "jogs" in the trend, as there are for I_1.
- χ decreases steadily down a family, again paralleling the trend in I_1.
- There is a rough equality of the electronegativities of an element and the element one column to the right and one row below it. Thus the electronegativities of Cl and O are 3.2 and 3.4, respectively. Those of N and S are also quite similar. This results because the increase in χ due to the rightward move and the decrease in χ due to the downward move roughly cancel.

Electronegativity is the primary concept on which we base our understanding of electron distributions among atoms in a molecule.

Bond Dipoles and Dipole Moment

The distribution of electrons in a molecule determines its shape. One frequent consequence of shape is that the centers of positive and negative charge in the molecule are displaced from one another, so that the molecule has a positive end and a negative end. Such molecules are said to possess total molecular dipole moments. It is important to be able to assess whether or not a molecule is polar; that is, whether it has a total molecular dipole moment. In this section, we will take some time to discuss the origins of total molecular dipole moments so that you will be equipped to do this assessment.

An arrangement in which a positive charge q+ and an equal negative charge q- are separated by a fixed distance, r, is called an electric dipole. The electric dipole is a vector quantity that is by convention taken to point from the positive to the negative end of the dipole. The magnitude (length) of the vector is the product of the charge, q+, and the distance between the positive and negative centers:

$$(3\text{-}3\text{-}1):\ \mu_d = \theta r$$

The electric dipole has units of charge times distance. For a dipole consisting of a charge of 1 electronic charge unit separated from an equal and opposite charge by a distance of 1×10^{-10} m, the dipole moment has the value 1.6×10^{-29} C-m.

The total molecular dipole moment of a molecule is in general the vector sum of various local electric dipoles that exist in the molecule. These local electric dipoles are of two types that we will call bond dipoles and lone pair dipoles. Local bond dipoles are generally larger in magnitude, hence more important, than lone pair dipoles.

They are thus the major contributors to the total molecular dipole moment, and we will discuss them first. A bond between two atoms, X and Y, in which the bond pair is unequally shared is called a polar bond. A polar bond gives rise to a local bond dipole. Unequal sharing occurs when X and Y have different electronegativities; the shared electron pair is on average closer to the more electronegative atom.

Consequently, the local bond dipole points along the bond, from the less electronegative atom (positive end) to the more electronegative atom (negative end). Assessment of the polarity of a bond is simple: if the two atoms involved in the bond are different either in identity or environment, the bond is polar. Only bonds between identical atoms in identical environments are non-polar. A heteronuclear diatomic molecule such as HCl provides an example of a local bond dipole. Heteronuclear means that the two atoms of the molecule are different. Chlorine is more electronegative than hydrogen, so the bond electrons between them are closer to Cl than to H. This makes the Cl atom slightly negative and the hydrogen atom slightly positive. The H-Cl bond is polar, with the electrons biased towards Cl. This polarity gives rise to a local bond dipole that points along the bond from H (positive) to Cl (negative). The second type of local electric dipole is the lone pair dipole, which exists at any atom having one or more lone electron pairs. A local lone pair dipole points, as expected, from the nucleus of the atom (positive) to the lone pair (negative).

Although strictly speaking there is a lone pair dipole generated by each lone pair in a molecule, the major contribution to the total molecular dipole is made by lone pairs on the central atom. Consequently, in examining a molecule for local lone pair dipoles, we will consider only lone pairs on the central atom, ignoring those on terminal atoms.

In addition to its local bond dipole, HCl contains three local lone pair dipoles, one associated with each of the lone pairs on the chlorine atom. The total molecular dipole moment of a molecule is the vector sum of all local electric dipoles (bond dipoles and central atom lone pair dipoles). The total molecular dipole moment of HCl is 0.36×10^{-29} C-m. This is due primarily to the bond dipole; the lone pair dipoles make a minor contribution, as is generally true. Assessment of the total molecular dipole of HCl is quite simple, because there are only two atoms.

For larger molecules, we will use the following systematic approach:

- First, examine each bond in the molecule for polarity. For each polar bond, draw a vector along the bond from the positive atom to the negative atom to represent the local bond dipole.
- Combine the local bond dipoles using vector addition methods to obtain the net bond dipole vector. The net bond dipole will usually be the major contributor to the total molecular dipole moment.
- Next, look for local lone pair dipoles on the central atom. Draw a vector from the atom symbol to each lone pair on the central atom. These are the local lone pair dipole vectors.
- Combine the central atom local lone pair dipoles using vector addition methods to obtain the net lone pair dipole vector.
- Combine the net bond dipole vectorially with the net lone pair dipole. The result is the total molecular dipole moment vector.

To illustrate the system, we consider the water molecule, which according to spectroscopy (and VSEPR) is bent, with an angle of 104°. Oxygen is more electronegative than hydrogen, so the O-H bonds are polar with the electrons closer to the oxygen than to the hydrogen atom.

Each O-H bond in the water molecule therefore has a local bond dipole that points from H to O. The bond dipoles are along with their vector addition to produce the net bond dipole. Because the water molecule is bent, the local bond dipoles add together to produce a non-zero net bond dipole that points from the center of the (imaginary) line connecting the hydrogen atoms (positive) to the oxygen atom (negative). Next, we find two lone pairs on the central oxygen atom. Each lone pair generates a lone pair dipole pointing from the oxygen nucleus to the lone pair.

The result is a non-zero net lone pair dipole that points from the oxygen atom (positive) to the center of the (imaginary) line connecting the lone pairs (negative). The net lone pair dipole and the net bond dipole point in the same direction. is the vector addition of the net bond and lone pair dipoles to produce the total molecular dipole moment of water. The magnitude of the total molecular dipole moment is 0.615×10^{-29} C-m. This substantial dipole is a major factor in the ability of water to dissolve a variety of substances, notably including ionic compounds.

Although bonds between different atoms are polar to at least some extent, as we saw above, many highly symmetrical molecules have no net dipole moments due to exact cancellation of the bond and lone pair dipoles. An example is the CO_2 molecule, which is linear. Each C-O bond is polar. Each therefore has a local bond dipole pointing directly along the bond from carbon (less electronegative) to oxygen (more electronegative).

However, the two equal local bond dipoles point in opposite directions and cancel exactly. The central carbon atom has no lone pairs, so no lone pair dipoles need be considered. Thus the molecule as a whole has coincident centers of + and - charge. We will see later that dipole moments contribute to the attractive forces that molecules exert on one another. To understand the magnitudes of these forces, it is essential that we be able to determine whether or not a molecule possesses a total molecular dipole moment.

Example: Which of the following molecules have (total molecular) dipole moments? SO_3, O_3, CCl_4, NH_3, PF_5.

Solution: Molecular polarity is assessed from molecular shape, which is in turn assessed from the Lewis structure. The method developed earlier leads to the following Lewis structures and shapes for these species. Lone pairs on the terminal atoms are not shown.

- *SO_3*: Each S-O bond is polar, with a local vector bond dipole pointing from S (less electronegative) to O (more electronegative.

All S-O bonds are the same when resonance is taken into account. The three local bond dipoles thus point to the corners of an equilateral triangle. When added vectorially, the result is zero. There are no lone pairs on the central S atom, so local lone pair dipoles need not be considered. There is no net molecular dipole; SO_3 is non-polar.

- *O_3*: Although the bonds in this molecule connect pairs of oxygen atoms, the two oxygen atoms of a bond are not in identical environments. Considering the bond between oxygens 1 and 2, we see that oxygen 1 is bonded only to oxygen 2, whereas oxygen 2 is bonded to oxygen 1, but in addition to oxygen 3. Oxygens 1 and 2 are in different environments, so the bond between them is polar. The difference between oxygens 1 and 2 is evident from their formal charges, which are 0 and 1+ in the resonance form shown.

The formal charges also enable us to assign the direction of the local bond dipole, which points from oxygen 2 to oxygen 1. The molecule is bent, so the two identical local bond dipoles do not add to zero. They cancel horizontally, but not vertically. The local lone pair dipole pointing from oxygen atom 2 to its lone pair only partially cancels the vertical net bond dipole (recall our statement that in general bond dipoles are more important than lone pair dipoles). Ozone has a net molecular dipole running from the center oxygen through the center of the molecular angle.

- *CCl_4*: From the Lewis structure for CCl_4, a perfect tetrahedral shape is expected (like methane, CH_4). The local bond dipoles are non-zero, with the negative end towards Cl, but cancel when added vectorially. The central carbon atom has no lone pairs. There is no molecular dipole moment. The cancellation of local bond dipoles may not be obvious here because of the three-dimensionality of the figure. However, the following argument may be used. Molecules having total molecular dipole moments are inherently less symmetric than molecules with zero molecular dipole moments.

For example, if you were to orient the ammonia molecule with the nitrogen lone pair pointing down rather than up, it is easy to tell that the molecule has been moved. In contrast, if you were to orient carbon tetrachloride with any of the three other chlorine atoms at the top, the molecule would look exactly the same as it does above; you would not be able to tell that it had been moved! Thus it has no dipole moment.

- *NH_3*: NH_3 has a trigonal pyramidal shape. The bonds are polar, with the negative end of the local bond dipole towards nitrogen. These will combine to give a non-zero net bond dipole colinear with an axis passing through the nitrogen atom and the center of the equilateral triangle defined by the H atoms. This is reinforced by the nitrogen local lone pair dipole. The molecule therefore has a molecular dipole moment with the nitrogen end of the dipole negative.
- *PF_5*: PF_5 is trigonal bipyramidal. The bonds are very polar because F is the most electronegative element. However, the two axial local bond dipoles oppose and cancel, and cancellation occurs in the equatorial plane as well, exactly as in SO_3 above (you should use trigonometry to prove this to yourself). There are no lone pairs on the central atom. There is no molecular dipole moment.

Isovalent Fragment to Structure

An approach to molecular structure based on analogical thinking. An oxygen atom has 6 valence electrons, so it requires 2 to achieve the octet. Suppose it gains 1 electron by forming a bond to hydrogen, giving the species H-O*. Here * represents the remaining odd electron on the oxygen atom that must pair with a second electron in bond formation. Because the species H-O* requires only 1 electron more, it is analogous to a halogen atom, for example, Cl*.

Again, * represents the seventh chlorine electron, the one that is shown unpaired in the dot structure for the atom. Since the O atom in H-O* is capable of forming one bond, like Cl*, OH* and Cl* are isovalent (iso = same, so same valence).

Further, we will call them fragments because they do not yet have electron octets. Finally, they are 1-bond fragments because they need form only one additional chemical bond. Combining terms, we will refer to HO* and Cl* as 1-bond isovalent fragments. To emphasize the 1-bond aspect of these fragments, we will write them as OH- and Cl-, where the dash represents an incipient bond, not a negative charge.

You should convince yourself that the following species are also 1-bond fragments, isovalent with HO- and Cl-:

- H_2N–
- H_3C–

- Br–
- H_2P–
- H_3Si–
- H–

Note that the number of atoms attached to the central atom in these fragments increases by 1 for each 1-unit decrease in the group number of the central atom.

Compatible fragments can be united to give the central atom of each fragment the octet. Thus two 1-bond fragments can be united. We therefore expect the existence of molecules in which two of the fragments from those listed above are combined.

In fact, all possible combinations of the 1-bond fragments above actually exist and are (within the framework of the fragment idea) related, or analogous.

Table: *The following structures are illustrative:*

Cl-Cl	HO-OH	HO-Cl	H_2N-OH	H_3C-OH
Chlorine	Hydrogen Peroxide	Hypochlorous Acid	Hydroxyl Amine	Methyl Alcohol

In similar fashion, the oxygen atom itself can be viewed as a 2-bond fragment, represented O=, because by forming 2 bonds it can achieve the octet.

The following species are 2-bond fragments isoelectronic with O=:

- HN=
- H_2C=
- S=
- $(CH_3)_2C$=

2-bond fragments are compatible; thus two of them can be united to give the central atom of each fragment the octet. The following molecules can be viewed as the union of two 2-bond fragments:

O=O	Dioxygen
$H_2C=CH_2$	Ethylene
$CH_2=O$	Formaldehyde
$(CH_3)_2C=O$	Acetone

Further, a 2-bond fragment is compatible (can unite) with two 1- bond fragments, as in the following molecules:

- OH_2
- NH_3 (HN= plus 2H-)
- CH_4 (H_2C= plus 2H-)

Finally, the nitrogen atom is a 3-bond fragment, requiring 3 electrons to achieve the octet. It may be represented as N(3).

Isoelectronic 3-bond fragments are:

- HC(3)
- (C(3))–
- $(O(3))^+$
- P(3)

Combination of such fragments gives the following molecules or ions:

- HC(3)CH acetylene
- C(3)O carbon monoxide
- N(3)N dinitrogen
- $N(3)O^+$ nitrosyl cation

A 3-bond fragment is compatible with three 1-bond fragments. You should generate some molecular formulas by combining 3-bond and 1-bond fragments. A single 2-bond fragment may be viewed as equivalent to two 1-bond fragments (X= equivalent to 2 Y-), and may replace them in a formula. Thus phosgene, $COCl_2$, may be "generated" from carbon tetrachloride, CCl_4, by replacing two 1-bond Cl fragments with the 2-bond fragment, O=. CO_2 follows from phosgene by a similar replacement of the remaining Cl- fragments.

Similarly, X(3) is equivalent to 3 Y-. It is quite easy to go from a known Lewis structure to several others using this analogical approach. The generation of a family of sulfur compounds. Successive replacement of two 1-bond F fragments with the 2-bond fragment, O=, leads to sulfur trioxide via two oxofluorides. Replacement of O= by another 2-bond fragment, HN=, leads to the Lewis structure for $S(NH)F_4$. Finally, replacement of three 1-bond F with one 3-bond fragment, N, generates the final molecule, SF_3N.

Importance of Molecular Stereochemistry

The importance of molecular shape cannot be overstated. It is the shape of the simple water molecule that determines its physical and chemical properties, and in particular makes it suitable for sustaining life. Shape has far reaching consequences in chemistry, biology, and the technology of materials. It is probably not an exaggeration to say that virtually all important biological processes are grounded in molecular shape, and, in particular, the ability of molecules to recognize one another based on shape. A familiar and extremely important example is found in the the double helical structure of the DNA molecule, responsible for the transcription of genetic information in all living organisms. The double helix is maintained because molecules recognize one another's shapes.

The guanine and cytosine molecules engage in three attractive interactions that hold them in specific positions relative to one another. In biochemical jargon, this is called a base-pairing interaction; it occurs in the gap between the two DNA strands. In 3-21b, adenine and thymine are shown in their base-pairing engagement, this time involving two shape-specific interactions. Clearly, molecular shape plays a crucial role in enabling these life-sustaining interactions. Chemists are actively conducting research in the area of molecular recognition; that is, the manner in which molecules recognize each other based on shape. This area of research promises the development of new and useful polymers and plastics; and mechanical and electrical devices that can be used in the engineering of extremely small (nanoscale) machines and computers.

It will also help to clarify the details of many biochemical processes that are currently only vaguely understood. Among these are the interaction(s) between enzymes and their substrates that lead to catalysis of substrate conversion to products; the role of retinal in vision; the mechanism of smell; and the mechanism of nerve transmission. We will have more to say about the nature of interactions between molecules. For now, it is important only to realise that such interactions are shape-driven.

Example: Application of structural concepts to an important problem: Why is the amide linkage in proteins planar?

Solution: Proteins are polymers constructed from amino acid monomers. This mean that single amino acids are strung together like snap beads to create an extremely long molecule called a protein, or polypeptide. There are some 20 amino acids used in making proteins, all having the basic structure shown in Figure. In the figure, R is a generic representation for a one-bond fragment that is attached to the central carbon atom. R is different for each distinct amino acid. Two amino acids link via reaction 3-3-2 below, in which the nitrogen atom of the NH_2 group of one amino acid becomes bonded to the carbon atom that is doubly bonded to oxygen in the second amino acid.

The process is shown in structural terms in figure:

(3-5-1): NH_2-CHR-COOH+H_2N-CHR'COOH → H_2N-CHR-C(=O)-NH-CHR'COOH + H_2O

Because water is produced as a result of the linkage, the process is called a condensation reaction. The group of atoms shown in Figure, consisting of the carbon and oxygen of the left amino acid and the nitrogen, hydrogen, and central carbon of the right amino acid, is collectively referred to as the peptide link. It is found experimentally that all of these atoms lie in the same plane—*i.e.*, the peptide link is planar. It is our goal to provide a rationale for this observation in terms of the concepts of Lewis structure and VSEPR.

An acceptable Lewis structure for the peptide link is shown in Figure. This is the structure arrived at by applying the systematic procedure, and minimizing formal charges. The structure places three electron groups around the peptide carbon atom. All of these groups are used to bond atoms, so the stereochemistry at this carbon atom is trigonal planar. The structure places 4 electron groups around the nitrogen atom, one of them a lone pair.

This leads us to expect tetrahedral electron group distribution, and trigonal pyramidal stereochemistry at the nitrogen atom. However, this stereochemistry at the nitrogen atom would prevent some of the atoms in the peptide grouping from lying in the same plane with carbons 1 and 2 and the nitrogen atom. In particular, C3 and the proton attached to N could not simultaneously lie in the C1-C2-N plane in any arrangement of the molecule. How then do we explain planarity?

With some thought, we realise that there is another acceptable resonance form involving the N, C2, and O atoms of the peptide link. This is obtained by simultaneously moving the N lone pair in to form a double bond with C2, while moving the double bond pair between C2 and O out onto the oxygen atom as a lone pair.

This form, shown in Figure, is less acceptable than that in d because it places formal charges of 1+ and 1– on N and O, respectively. However, these are not unusual formal charges for these atoms, and they are not in conflict with electronegativities. The most important aspect of the structure in 3-22e is that there are now 3 electron groups around the nitrogen atom! Judging from this resonance form, trigonal planar stereochemistry is expected at N, and the planarity of the atoms C2, N, H, and C3 is required. Further, the double bond between N and C2 also requires C1 and O to be in this same plane. The planarity of the peptide link is understandable. The key question is this: when two resonance forms place different numbers of electron groups around an atom, what is the stereochemistry at that atom?

Based on our earlier discussion of resonance, we know that the molecule is not rapidly switching back and forth between the two forms. Rather, there is a single molecular structure that may be considered as an average of the resonance forms. And, in point of fact, the actual stereochemistry at N is trigonal planar. We will find the following rule to be useful in all situations like this:

When the reasonable resonance forms for a molecule place different numbers of groups around a particular atom, the stereochemistry at that atom is consistent with the smallest number of electron groups in any resonance form.

Our final example teaches us an extremely important lesson. The principles of structure and stereochemistry that we developed from and applied to small molecules apply equally as well for large molecules. Thus the carbon, nitrogen, oxygen, and sulfur atoms in proteins follow the same bonding and stereochemical rules that apply in small molecules. In fact, the behaviour and function of biomolecules is a consequence of the adherence to these rules.

Chapter

2

Intramolecular Chemistry

The electrical force which is responsible for all attractive interactions between atoms, molecules, and ions. The electrical force is described quantitatively by Coulomb's Law. A system in which forces act can possess two types of energy. These are kinetic energy, KE, and potential energy, PE. The magnitude of the kinetic energy is a function only of temperature, as we have seen. Potential energy arises from the action of the Coulomb force. We will discuss the various ways in which Coulomb forces operate in atomic/molecular systems, manifesting in both intramolecular and intermolecular contexts. Intramolecular forces include the forces between atoms in covalent molecules; the forces between oppositely charged ions in ionic compounds; and the forces between identical atoms in metals. Intermolecular forces include the forces that operate between one atom or molecule and another, or between the ion of one substance and the molecules of another. Although there are differences among the various types of forces, they are all fundamentally similar: they give rise to potential energy between the interacting particles, described by the potential well. We shall see that an interplay between the depth of this well and the thermal kinetic energies of particles is responsible for many of the physical properties of substances.

Types of Energy

An atom or a collection of atoms can possess energy of only two types. Kinetic energy (abbreviated KE) is energy of motion. A body of mass, m, moving with velocity, v, has kinetic energy given by equation.

$$KE = mv^2/2$$

When mass is expressed in kg and velocity in m/s, KE has units of kg m^2/s^2, or Joules (J). One Joule is defined as 1 kg m^2/s^2. The kinetic energy of atoms and molecules depends only on temperature, through the simple expression in 6-1-2.

$$\text{Average KE} = 3kT/2$$

Here Average KE is the average kinetic energy of a single atom or molecule; k is Boltzmann's constant; and T is the Kelvin temperature. Potential Energy (abbreviated PE) is energy of position. A rock poised on top of a hill, even though it is not moving, has the potential to move if given a small push. It therefore has potential energy. This potential energy is converted to kinetic energy as the rock rolls down the hill. All energy is either kinetic or potential. The total energy of any mechanical system is the sum of its kinetic and potential energies:

$$E = KE + PE$$

Potential energy is an abstract concept; it is difficult to understand how something can possess energy simply by virtue of its position. A weight rests on a platform at height h above the ground. The weight is connected by a rope, across a pulley, to another platform that carries a load. If we remove the weight from the platform, it will fall, and in the process will lift the loaded platform. But in lifting the load a distance h, the weight has done work (a form of energy) on the load. To do work, the weight must use energy.

It possesses the necessary energy to do the work by virtue of its position at distance h above th earth. Thus the concept of potential energy presupposes a force of attraction or repulsion. The weight falls to earth because there is a gravitational force of attraction between the earth and the weight. To move the weight from the ground to the platform initially, we would have had to use an amount of energy sufficient to overcome this force of attraction. The amount of energy used to lift the weight is stored in the earth-weight system as potential energy. We retrieve this energy by allowing the earth and weight to come together. Potential energy is a consequence of a force.

Our example has been based on the gravitational force because this is a familiar situation. There are 3 forces other than gravitation. One of these is the electromagnetic force. This is the force by which electrons are attracted and held by the nucleus in an atom. The electromagnetic force is responsible for all interactions between atoms and molecules. It is far stronger than the gravitational force, which may be neglected in discussing atomic and molecular interactions (the electromagnetic force of repulsion between two electrons is 10^{42} times stronger than the gravitational force of attraction between them!).

The electromagnetic force operating at the atomic level is responsible for:

- The attraction of electrons to the atomic nucleus;
- The bonding together of atoms in a molecule;
- The attraction between oppositely charged ions in an ionic compound;
- The attraction between molecules that leads to liquefaction and solidification;
- The flow of electrical current;
- Your inability to push your finger or hand through a wall.

The magnitude of the electrical force is given by Coulomb's Law, equation.

$$F = q_1q_2/W^*r^2$$

q_1 and q_2 are the values of the two interacting electrical charges, in Coulombs, and r is the distance between them, in m. W is a constant that adjusts the units. Its value is 1.113×10^{-10} $C^2N^{-1}m^{-2}$. The force described by Coulomb's Law is responsible for all attractive forces between atoms, molecules, and ions. The potential energy associated with the Coulomb force is called the Coulomb potential energy. Because force is the negative derivative of potential energy with respect to distance (F = -dPE/dr), the Coulomb potential energy can be obtained by integration of Coulomb's Law.

$$PE = q_1q_2/W^*r$$

When one of the charges is positive and the other negative, the potential energy is negative at all finite values of r.

Coulomb Forces and the Potential Well

Atoms, molecules, and ions exert attractive forces on one another because the atomic nuclei in one species attract the electrons in the other. These are called molecular forces, and the potential energy thatresults from the action of these forces is called molecular potential energy. We make several general statements about molecular forces and potential energy, and the relationship between them. The statements are equally valid for a pair of particles or a large collection of particles (here we are using the word, "particle", generically to represent an atom, a molecule, or an ion).

- Molecular forces are electrical;
- Molecular forces are attractive except at very close distances, and described by Coulomb's Law;
- The magnitude (strength) of the force between 2 particles varies with their distance of separation. The force is zero for an infinite separation, increases as the particles get closer, becomes a maximum when the particles just touch, and becomes strongly repulsive when the centers of the particles are separated by less than one particle diameter, due to interpenetration and repulsion of electron clouds. The force has two components: one operates only at very small distance and is repulsive; the other acts at all distances and is attractive. Thus

$$F = q_1q_2/W^*r^2 + B$$

B is a positive, repulsive force that operates only at very small distance.

- Action of the molecular forces draws the particles together (tends to cause them to aggregate) and results in a decrease in the potential energy of the pair (or collection) of particles. The force becomes stronger and the potential energy less until the particles touch. At distances less than 1 particle diameter, the force is repulsive and PE increases again. For all types of molecular forces, potential energy varies with distance of molecular separation, r. The plot is shaped like a valley, or well. It is usually called a potential energy well. The depth of the well is the energy that must be added to increase the distance of separation between particles from r_1 to r_3 or greater.
- There are several types of molecular forces, which we will discuss shortly. All of them are characterized by a potential well of similar shape. However, the depth of the well increases with the strength of the forces.

Region: The particles, shown as two small circles just above the horizontal axis, are separated by large distances. Taking 2×10^{-10} m as a typical particle radius, the distances of separation in region 1 are 10 or more times larger. At large separation distance, molecular forces are nearly zero. For convenience, we take the potential energy to be zero when the forces are zero, so the PE curve coincides with the horizontal axis at large r. This situation of very large separation between atoms, molecules, or ions is a state of *deaggregation*. The gas phase of a pure substance is an example of a deaggregated state.

Region: The particles have approached closely enough to exert noticeable forces on one another. Action of the forces lowers the potential enrgy. The closer the particles approach (*i.e.*, the more aggregated they become), the stronger the forces and the lower the PE. The particles just touch. At this point, attractive forces are maximized and potential energy is at a minimum. This is an aggregated state. The solid phase of a pure substance is an example of an aggregated state.

Region: PE increases very rapidly as r decreases. In this region, the somewhat deformable electron clouds of the particles are being pushed into one another. Since electrons repel one another, two particles that have interpenetrated exert a repulsive force on one another, and tend to fly apart. Region shows that when particles are forced together so that their centers are separated by less than one molecular diameter, repulsive forces, hence PE, rise very rapidly. This is the force that keeps you from falling through the floor.

The attractive forces are maximized and the PE minimized when the atoms, molecules, or ions just touch, at the bottom of the well. The distance corresponding to the potential energy minimum is the distance of closest approach. We have called this the aggregated state. In an ionic compound, such as KBr, this corresponds to a situation in which positive ions are surrounded by negative ions, and vice versa.

In a covalent compound, such as H_2O, the bottom of the well represents the situation in which the hydrogen atoms are covalently bonded to the oxygen atom, the bonds are separated by an angle of 104 °, and the distance between the H and O atoms corresponds to r at the bottom of the well. A collection of He atoms or methane molecules can maximize the contact between themselves by aggregating such that a particular molecule touches as many other molecules as possible. This organization is referred to as close packing, and characterizes the solid state of a pure substance. PE is minimized in the aggregated state. Thus shows how potential energy changes as a substance is converted from the deaggregated to the aggregated state or vice versa. Notice that potential energy increases when an aggregated state is converted to a deaggregated one:

$$\text{Aggregated} \rightarrow \text{Deaggregated DPE} > 0$$

This is an idea that has broad applicability; we will encounter it many times.

The phases (solid, liquid, and gas) of a pure substance fit the aggregation/deaggregation model very well. In the solid phase, the fundamental units of a substance are packed together in an orderly arrangement that minimizes potential energy. This is the aggregated state, in which intermolecular attractive forces are optimized. In the gas phase, the fundamental units are widely separated and experience only very weak forces from one another.

This is the deaggregated state. We might ask where the liquid state is on the diagram. From macroscopic observation, we know that the molar volume of a substance, *e.g.* water, in the liquid phase is very little different from the molar volume of its solid phase. These two aggregated (or condensed) phases have molar volume roughly 1000 times smaller than that of the gas phase. The liquid phase has distance of molecular separation and potential energy only somewhat greater than those of the solid, at the point labelled r_2.

The condensed (aggregated) phases of matter and the conversions between the gas, liquid, and solid phases. Prior to that, it is important that you have an understanding of the types of forces that operate within and between molecules, and their relative strengths. This will enable you to understand, for example, why some substances exist as high-melting solids under normal conditions, while others exist as low melting solids, liquids, or gases.

Types of Molecular Forces

The nature and structure of a substance determine the type and strength of molecular forces that operate within it. We divide such forces into two broad categories. Intramolecular forces operate within the molecules or fundamental units of a substance. These are the covalent bonds within a molecule of a molecular substance, the forces between ions in an ionic compound, and the forces between atoms in a metal. Intermolecular forces operate between, rather than within, the molecules of a covalent substance; the atoms of a monatomic element; or the ions of one substance and the molecules of another.

As a rule, intramolecular forces are much stronger than intermolecular forces. Both types of force, however, give rise to potential energy and a potential well. A listing of the important forces, arranged within these categories, follows:

Intramolecular Forces

- Attraction between positive and negative ions in a crystal of an ionic compound.

- Covalent bonds linking atoms in a network structure, such as that of silicon dioxide (sand).
- Covalent bonds in molecular substances.
- Metallic bonds

Intermolecular Forces

- Attractions exerted by one molecule of a molecular substance on another, such as the force of attraction between water molecules in ice.
- Attractions between atoms of the noble gas elements, helium through radon.
- Attractions between molecules of one substance and molecules of another, as when two liquids are mixed, or a molecular solid such as sugar is dissolved in a liquid.
- Attraction between molecules of one substance and ions of another, as when an ionic compound dissolves in a liquid.

In succeeding sections, we will discuss each of these molecular forces, and their relative magnitudes. We will begin with intramolecular forces.

Intramolecular Forces

The Ionic Bond

When two elements of widely different electronegativities combine, the result is an ionic compound. An example of an ionic compound is potassium bromide, KBr, in which the electronegativity difference between bromine, in group 17, and potassium, in group 1, is 2.14. Ionic bond formation involves transfer of an electron from the potassium atom, with its single valence electron and core charge, Z_{core}, of 1 unit, to the bromine atom, with its seven valence electrons and large Z_{core} of 7 units. The electron transfer generates a full valence shell for each species. The ionic bond can be understood in terms of three energy terms. To remove an electron from the potassium atom to generate the potassium ion requires input of the first ionization energy of potassium. The process is shown in 6-4-1.

$$(6\text{-}4\text{-}1): K \rightarrow K^+ + e^- \quad I_1 = 420 \text{ kJ/mole}$$

I_1 for potassium is relatively small due to the minimum core charge found for group 1 elements. To add an electron to the bromine atom to form the bromide ion releases an amount of energy equal to the first electron affinity of bromine.

$$Br + e^- \rightarrow Br^- \quad -A_1 = -325 \text{ kJ/mole}$$

The first electron affinity for bromine is substantial, because the large core charge of the bromine atom exerts a reasonable force on the extra electron. Despite this, to accomplish the electron transfer from potassium to bromine requires that 420 – 325 = 95 kJ/mole of energy be supplied. In other words, the electron transfer is substantially unfavorable! Thus we find that, contrary to a commonly held view of ionic bond formation, potassium does NOT "like" to transfer an electron to bromine; the process is unfavorable, and is not responsible for the strength of the ionic bond. What, then, is?

Generation of positive and negative ions results in a strong attraction between them, according to Coulomb's Law. A single positive and negative ion are shown coming together to form an ion pair in reaction below.

$$K^+ + Br^- \rightarrow K^+Br^- \text{ (ion pair)}$$

Formation of one mole of ion pairs releases 589 kJ of energy, much more than was required to generate the ions. We conclude that it is the Coulomb attraction between oppositely charged ions that drives the formation of the ionic bond; this happens despite the need to form ions, which is unfavorable. In fact, ionic compounds do not consist of ion pairs of the type shown in reaction above. Rather, they consist of extended, highly organized aggregates in which positive and negative ions alternate. Figureshows both the ion pair and the lattice of potassium bromide.

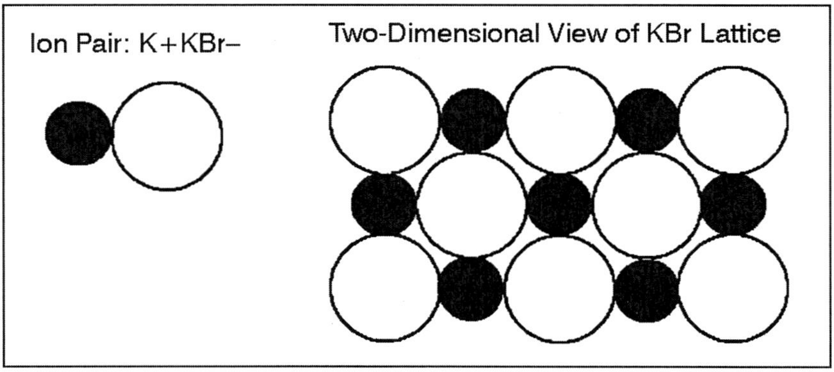

***Figure:** Potassium Bromide, Ion Pair and Lattice*

The lattice is more stable than a collection of ion pairs because it surrounds each ion with ions of the opposite charge. This maximizes the attractive forces between positive and negative ions. In the ion pair, the positive charge of the potassium ion is offset on one side by the bromide ion, but is still exposed on all other sides.

This exposed positive charge attracts bromide ions of other ion pairs, causing the ion pairs to condense to a lattice. Formation of a lattice containing 1 mole each of potassium and bromide ions as in reaction releases 672 kJ of energy. This number, about 50% larger than the amount released in formation of the equivalent amount of ion pairs, represents the depth of the potential well for potassium bromide.

$$K^+(g) + Br^-(g) \rightarrow KBr \text{ (lattice)}$$

The energy change that occurs when 1 mole of ionic substance in the solid lattie is formed from separated positive and negative ions in the gas phase is called the lattice energy, U, of the ionic compound. The lattice energy is given by an expression having the form of:

$$U = KZ^+Z^-/r \text{ m}$$

Here K is a collection of physical constants, Z^+ is the cation charge, Z^- is the anion charge, and r is the distance of separation of the cation and anion in the lattice. Because the charge of the anion, Z^-, is negative, the lattice energy is a negative quantity, as expected. This means simply that the solid lattice is more stable than the separated gas phase ions. Note that this equation is essentially the expression for Coulomb potential energy. Without doing any calculations at all, we can draw some important conclusions based on reaction.

- The potential well for an ionic compound of a 2+ cation with a 2– anion will be about 4 times deeper than that for a 1+/ 1– ionic compound, like KBr, due to the product of charges in the numerator of equation. Knowing the lattice energy for KBr allows us to estimate the lattice energy for MgO: about -2600 kJ/mole
- The potential well for an ionic compound will be deeper when cation and anion are small than when they are large. Small cation and anion means small distance of separation, r, leading to a more negative U. The lattice energy for LiF is substantially more negative than that for KBr, despite the fact that both are 1+/1– compounds, because Li+ and F- are smaller ions.

Lattice energies for a number of common ionic compounds are given in Table.

***Table:** Lattice Energies of Ionic Compounds*

Charge Type	***Substance***	***U, kJ/mole***
1+/1–	LiF	–1009
	NaCl	–774

Contd...

Charge Type	Substance	U, kJ/mole
	KBr	–672
	CsI	–584
2+/2–	MgO	–2585
	BaO	–1978

Example: The lattice energy for NaCl is –774 kJ/mole. The radii of the sodium and chloride ions are 116 pm and 167 pm respectively. Estimate the lattice energy for CsI, which has ion radii of 181 and 206 pm.

Solution: NaCl and CsI are both 1+/1- ionic compounds. They should differ in lattice energy only through the r term of 6-4-5. Because the distance of separation is larger in CsI, the lattice energy should be less negative. We assume that r can be approximated by adding the radii of the ions. Thus for NaCl, r = 283 pm; for CsI, r = 387 pm.

The lattice energies should be inversely related to the distance of separation of the ions:

- U(CsI)/U(NaCl) = r(NaCl)/r(CsI)
- U(CsI)/(–774) = 283/387
- U(CsI) = –570 kJ/mole

This compares reasonably well with the value in Table.

Example: Arrange the following ionic compounds in order of increasing intramolecular forces: KI, LiCl, MgO, CaO, Al_2O_3.

Solution: KI and LiCl are 1+/1- ionic compounds. MgO and CaO are 2+/2- ionic compounds. Finally, Al_2O_3 is a 3+/2- compound. Based on ion charges and equation, we can arrange these in three categories:

1. KI, LiCl
2. MgO, CaO
3. Al_2O_3

The 1+/1– compounds have the least negative lattice energies. That for Al_2O_3 should be most negative, because the ion charges are largest. Within each category, the compound with the largest ions has the least negative lattice energy.

This reasoning leads to the following order:

$$U(KI) > U(LiCl) > U(CaO) > U(MgO) > U(Al_2O_3)$$

Here the greater than sign (>) is applied in the algebraic sense, in which numbers are considered to become smaller as they get more negative. In this convention, -50 is greater than -100.

There are very few compounds that are legitimately considered to be ionic; that is, in which electron transfer from metal to non-metal is complete. They are limited to compounds of the group 1 metals (Li, Na, K, Rb, Cs) with fluorine, chlorine, and oxygen. Compounds of the group 2 metals with oxygen, sulfur, and the halogens form typical ionic lattices, and are usually considered to be ionic compounds. However, there is considerable covalent character in the bonding.

Based on our discussion of the ionic bond in this section, it is understandable why ionic compounds are invariably solids at room temperature, with high melting and boiling points. The strong, non-directional Coulomb attractions acting between oppositely charged ions in the lattice can be overcome only at very high temperature.

The Covalent Bond

Two atoms of relatively large and similar electron-egativities (*i.e.*, non-metals) tend to form covalent compounds rather than ionic compounds. In covalent compounds, atoms share rather than transfer electrons to achieve full valence shells, because both atoms have similar substantial values of the core charge. Large Z_{core} tends to make electron transfer unfavorable. The shared electrons are paired, and are usually considered to be localized between the bonded atoms. In this section, we will discuss the chemist's view of the covalent bond. Two separated chlorine atoms, each with 7 valence electrons distributed in the 3s and 3p valence orbitals.

Each Cl atom has a single odd electron in a 3p orbital. It is these electrons that will be paired and shared in the covalent bond. For the two atoms to bond by sharing electrons, it is necessary that they establish a mutual region in which the shared pair resides. They accomplish this by merging, or overlapping, the atomic orbitals containing the electrons to be paired. Because these atomic orbitals have distinct spatial directions, the bonded atoms must maintain a definite orientation relative to one another in order to preserve the overlap of orbitals. Thus covalent bonds are directional.

The operation of Coulomb's Law in the covalent bond is perhaps less obvious than in the ionic bond. The two chlorine nuclei and the two electrons of the covalent bond, and indicates the various Coulomb forces that operate in this collection of centers. There are several interactions: attraction of nucleus 1 for electron 1; attraction of nucleus 2 for electron 1; attraction of nucleus 1 for electron 2; attraction of nucleus 2 for electron 2; repulsion of nuclei 1 and 2; and repulsion of electrons 1 and 2. Note, however, that because the bonding electrons are required to be between the nuclei so that they can occupy the overlapping atomic orbitals, the nucleus-electron attractive forces are maximized, and the nuclear-nuclear repulsions are minimized, because the electrons shield the two nuclei from each other. Thus the strength of the covalent bond is attributable to the simultaneous attraction of both electrons to both nuclei.

The lengths and strengths of covalent bonds—the position and depth of the covalent bond potential well—vary with the bonded atoms. In general, the strengths are substantial and comparable with those of ionic bonds.

Thus to break all of the covalent bonds in a molecule of methane, CH_4, requires the input of 1662 kJ of energy. Methane lies in a deep potential well relative to the separated atoms. Generally speaking, the following generalizations hold for covalent bond strengths.

- Small atoms form relatively short bonds, and large atoms form relatively long bonds.
- Short bonds tend to be stronger than bonds formed by larger atoms because the bonding electrons are closer to the bonded nuclei. For example

Bond	***Length***	***Strength***
Cl-Cl	198 pm	240
Br-Br	228 pm	190
I-I	266 pm	149

- The word "tend" is emphasized here, because there are exceptions. Notably, single bonds of the small atoms, N, O, and F are very weak.
- Multiple bonds are stronger than single bonds. For example, the single, double, and triple bonds between carbon and oxygen have the following strengths:

  C-O 358 kJ/mole

  C=O 799 kJ/mole

  C(3)O 1071 kJ/mole

This seems reasonable, because there are more electrons localized between the nuclei and attracted by both. Strengths and lengths of some typical covalent bonds are presented in Table. A more complete set of covalent bond strengths is found in Table below.

Table: Strengths and Lengths of Covalent Bonds

Bond kJ/mole	***Bond Energy (Strength), Single Bonds***	***Bond Length, pm***
H-H	432	74.2
H-F	565	91.8
H-Cl	428	127.4
H-C	411	109
H-N	386	101
H-O	459	96
C-C	346	154
C-F	485	135
C-Cl	327	177
C-O	358	143
C-N	305	147
N-N	167	145
N-O	201	140
N-F	283	136
O-O	142	148
O-F	190	142
Multiple Bonds		
C=C	602	134
C=O	799	120
C=N	615	
N=N	418	125
N=O	607	121
O=O	494	120.7
S=S	425	188.7
C(3)C	835	120
C(3)O	1072	113
C(3)N	887	116
N(3)N	942	109.8
P(3)P	481	189
As(3)As	380	

Example: Arrange the following single bonds in order of increasing length. Then arrange them in order of increasing strength.

H-O, O-Cl, Cl-Cl, H-Cl

Solution: Longer bonds are formed by larger atoms; atoms become larger as we descend the periodic table.

The sizes of the atoms involved here are ordered as follows:

Cl > O > H

The bond lengths are then expected to be in this order:

H-O < H-Cl < O-Cl < Cl-Cl.

The strengths of bonds tend to decrease with increasing length.

Thus:

Cl-Cl < O-Cl < H-Cl < H-O.

Although simplified, this section gives us an appreciation for the origin and strengths of covalent bonds. We now turn to a problem with our view of the covalent bond having to do with molecular stereochemistry. The prediction and rationalization of molecular stereochemistry (shape): the VSEPR theory.

Thus VSEPR predicts the tetrahedral geometry for methane, in exact agreement with experiment. The interpretation of the covalent bond in terms of overlap of atomic orbitals of the bonded atoms requires that we understand molecular shape in terms of the distribution in space of the atomic orbitals of the central atom, which are presumably overlapped with suitable terminal atom orbitals in covalent bonds. We pose this question: are the observed shapes of molecules consistent with the distribution in space of the atomic orbitals of the bonded atoms? If we ask this question of methane, the answer must be NO.

On one hand, the C-H bonds are known to make 109.5^{o} angles with one another. On the other hand, the s and p valence orbitals of carbon have spatial orientations that are not consistent with this angle. The s orbital is spherical, hence non-directional; and the p orbitals make mutual angles of 90°. In fact, the bond angles in most molecules in which 3, 4, or 5 electron groups surround the central atom are inconsistent with this 90° orbital orientation. This presents a problem for our view of the covalent bond.

This problem was recognized in the 1920s, when the modern quantum theory was applied to the hydrogen atom and the spatial shapes of the atomic orbitals were discovered. Tetrahedral stereochemistry at carbon in methane had been proposed in 1874 by Vant Hoff on the basis of the number of isomers produced when hydrogen atoms were replaced by different atoms. At that time, though, people knew nothing of atomic orbitals. Thus the new quantum theory was faced almost immediately with a challenge: explain the shape of methane. This problem was tackled and resolved by Linus Pauling, who made during his life almost innumerable substantial contributions to science.

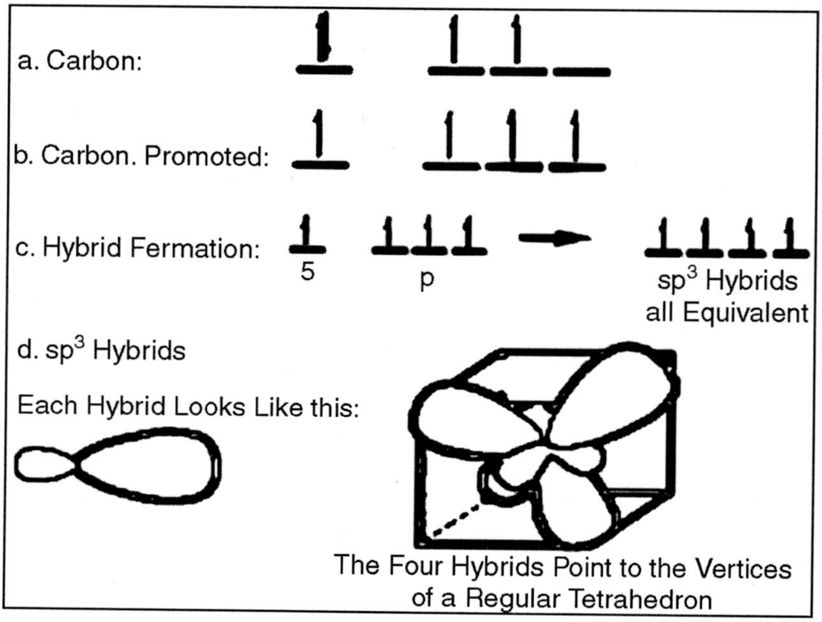

***Figure:** The Electron Configuration of Carbon*

Pauling proposed that the carbon atom uses hybrid orbitals to form four identical bonds to hydrogen atoms, directed to the vertices of a tetrahedron. His theory of hybrid atomic orbitals is still widely used today to rationalize (not predict) molecular stereochemistry. Applied to methane, the theory goes as follows.

In its most stable electron configuration, the carbon atom. The s orbital is fully occupied with an electron pair, and two of the three p orbitals are singly occupied. Based on this arrangement, we might expect carbon to form two bonds using the two odd electrons. However, this would fail to give it the octet. Thus we imagine that carbon moves, or "promotes", one of the 2s electrons to the empty p orbital. This prepares it to form 4 bonds. However, bond formation obviously does not occur using the pure atomic orbitals, because these can not produce a tetrahedral stereochemistry when overlapped with 1s orbitals of hydrogen atoms. Pauling proposed that the carbon atom creates from its 4 valence orbitals four new, equivalent orbitals called "hybrids."

The word hybrid means the same thing here as in agriculture: a blend or combination of two strains of a species. He called these sp^3 hybrids to indicate that they are composed of a mixture of a single s and three p orbitals.

Hybrid formation is shown schematically in Figure. Indeed, he showed mathematically that it is possible to combine the four s and p atomic orbitals to produce four new orbitals, and that these new orbitals are directed to the corners of a tetrahedron. The four hybrids are shown in Figure above. Pauling thus "squared" fact with quantum theory by proposing hybrid orbitals. Let's apply these ideas to a few more situations.

The phosgene ($COCl_2$) molecule is pictured in Figure. It has been experimentally determined that the molecule is trigonal planar (as expected from VSEPR), with a double bond between carbon and oxygen. How is the trigonal planar stereochemistry, involving 120° bond angles, to be rationalized with the 90° spatial distribution of carbon 2p orbitals? Pauling showed that the 120° bond angle results if one assumes that C hybridizes the s and two of the three p orbitals to produce sp^2 hybrids. Mathematically, the mixing of an s and two p's does indeed produce three equivalent new orbitals, directed to the corners of an equilateral triangle.

The plane in which the 3 hybrids lie depends upon which two p orbitals are used in the process. By convention, chemists normally choose to use px and py in the hybrids, which requires that the hybrids lie in the x-y plane of a Cartesian coordinate system. The pz orbital, not used in the hybrids, retains its usual shape and orientation along the z axis. Note, however, that it still contains an odd electron.

Once the hybrids are formed, each containing a single electron, carbon forms covalent bonds to the two hydrogen atoms and the oxygen atom. Each bond is formed by overlap of one carbon hybrid with one atomic orbital of the bonded atom (H or O). In forming these three bonds, orbitals are overlapped along the line joining the bonded nuclei. Bonds involving overlap of this type are called *sigma bonds*. At this point, bonding is not complete because carbon shares in only 7 electrons, as does oxygen.

Each can complete the octet by forming one more bond, using the p orbitals oriented along the z axis. Because these do not point at one another, but instead are parallel, overlap must occur in sideways fashion. Bonds formed by sideways overlap similar to this are called *pi bonds*. At this point, bonding is complete. We say that in phosgene, carbon uses sp^2 hybrids to sigma bond to H and O, and in addition forms a pi bond to O. The sp^2 hybrids are consistent with the trigonal planar sterochemistry.

We are ready to make a few useful generalizations about hybrid orbital theory. Having done that, we will look at an example.

- The hybridization of an atom cannot be determined unless its stereochemistry is known. Hybrid orbitals cannot be used as a predictor of stereochemistry.

  CH_4 is tetrahedral at carbon; therefore, carbon uses sp^3 hybrids.NOT carbon uses sp^3 hybrids; therefore CH_4 is tetrahedral at C.
- The number of hybrids obtained is equal to the number of atomic orbitals used to make them. Thus, for example, there must be three sp^2 hybrids because three atomic orbitals (one s and two p) are used.
- An atom uses hybrid orbitals for its sigma bonds and lone pairs. Thus the number of hybrids required is equal to the sum of the number of lone pairs and sigma bonds:

  4 sigma bonds + 0 lone pairs requires 4 hybrids, thus sp^3 $(1 + 3 = 4)$

  3 sigma bonds + 1 lone pair requires 4 hybrids, thus sp^3 $(1 + 3 = 4)$

  2 sigma bonds + 2 lone pairs requires 4 hybrids, thus sp^3 $(1 + 3 = 4)$

  3 sigma bonds + 0 lone pairs requires 3 hybrids, thus sp^2 $(1 + 2 = 3)$

  2 sigma bonds + 1 lone pair requires 3 hybrids, thus sp^2 $(1 + 2 = 3)$

  2 sigma bonds + 0 lone pairs requires 2 hybrids, thus sp $(1 + 1 = 2)$
- Unhybridized p orbitals are used for pi bonds. The number of pi bonds equals the number of unhybridized p orbitals.

The flow of logic in determining hybrids used by the central atom is as follows. First, count the number of electron groups around the central atom; then apply VSEPR to determine the stereochemistry at that atom. The number of hybrids is the same as the number of electron groups counted for VSEPR. A number of atomic orbitals equal to the required number of hybrids is used in forming the hybrids. The sum of the exponents in the hybrid designation indicates the number of hybrids of that type, and the number of atomic orbitals of each type used to make them.

Ionic and Covalent Bonding—A Continuum

The bond between two identical non-metal atoms, as in Cl_2, is purely covalent; that is, each atom has an exactly equal share in the bonding electron pair. Pure covalent bonds exist in the elemental forms of many of the non-metals (for example, all of the diatomic halogens of group 7; O_2 and S_8 of group 6; N_2 and P_4 of group 5; C and Si of group 4; and others). Bonds between different non-metal atoms are covalent, in that they involve electron sharing, but polar, meaning that they have some ionic character. The bonds between chlorine and the other elements of period 3 are shown below. The transition from pure covalent to polar covalent to ionic as we move from right to left in the period is clear.

Na-Cl — Mg-Cl — Al-Cl — Si-Cl —— P-Cl —— S-Cl —— Cl-Cl
ionic — ionic — ionic — polar cov—polar cov—polar cov—cov

Although the Na-Cl and Mg-Cl bonds are usually considered to be ionic, they are not purely ionic; electron transfer from the sodium or magnesium atom to the chlorine atom is not complete. There is no such thing as a pure ionic bond. The process of electron transfer from a metal atom to a non-metal atom generates a positive ion and a negative ion. The positive ion is electron deficient, and attracts an electron pair of the negative ion strongly. Thus the very act of electron transfer creates a disparity that must then be offset by partial electron sharing. If the anion is one of the very electronegative non-metals (fluorine or oxygen), it can resist the pull of the cation effectively, and hold on to the electron pair.

Anions other than F- and O^{2-}, however, cannot completely resist the attraction of the cation for their lone pairs. These atoms share a lone pair with the cation to form a bond that, although it is very polar, is nonetheless partially covalent. As a general rule, the higher the charges on the cation and anion, the more covalency in the bond. A highly charged cation (3+ or 4+) is small and compact, and exerts a powerful attractive force on electrons. A highly charged anion (2–, 3–) holds its extra electrons weakly and readily shares them to some extent with the cation. The following series of compounds illustrates nicely the evolution from largely ionic to covalent bonding as cation charge increases:

Compound	*Physical State, 298 K*	*Melting Point °C*
NaCl	Solid	801
$MgCl_2$	Solid	714
$AlCl_3$	Solid	190
$TiCl_4$	Liquid	–25

The decrease in melting point indicates increasing covalency in the bonding. Titanium tetrachloride, $TiCl_4$, is a liquid at room temperature, indicating that it is molecular, not ionic.

The Metallic Bond. Elements of similar but relatively low electronegativities (*i.e.*, metals) form metallic bonds. It is these bonds that are operative in a typical metal and that are responsible for metallic properties: reflectivity, conductivity, malleability, and strength. Metallic bonds form between atoms that have fewer valence electrons than they have valence orbitals. We shall try to develop a picture of the metallic bond by considering sodium metal, which exists as a very low melting, highly reactive solid at room temperature. Figure shows the Lewis dot symbols for two sodium atoms.

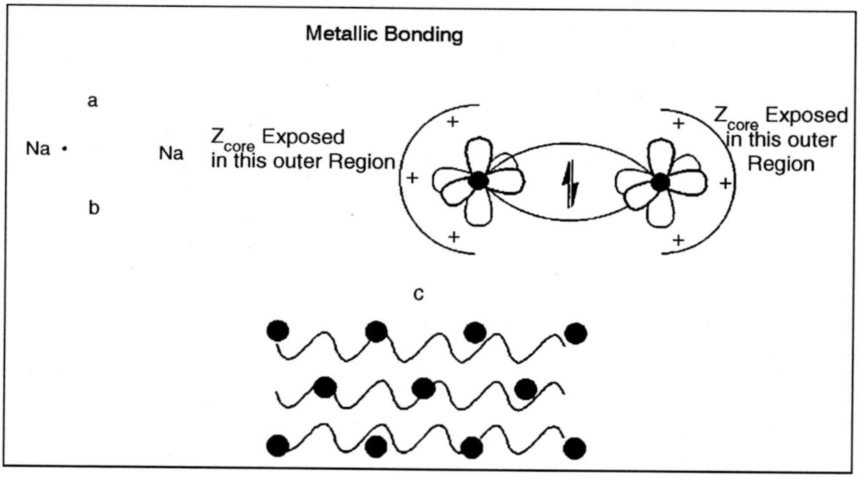

***Figure:** Metallic bonding*

The single odd electron is in the 3s valence atomic orbital of the sodium atom. At first glance, it might be expected that two sodium atoms, each with a single unpaired valence electron, would form a single covalent bond by overlapping the 3s valence orbitals and pairing the odd electrons in the resulting common region of space. There are problems with this bonding situation however. First, bonding of this type would give each sodium atom only a duet of electrons, nowhere near the required octet. Each atom would have three empty 3p orbitals protruding from it into space. Second, pairing the 3s valence electrons localizes them in the region between the two nuclei.

They no longer have probability of being found in the peripheral regions opposite the bond. Because these are the only valence electrons, the regions marked with + symbols in the figure are exposed to the full core charge (+1) of the sodium atom. (The + signs in the figure do not represent actual positive charges; instead, they are intended to indicate that another atom located in this region would experience the core charge of the sodium atom near it).

Here is the key point: another sodium atom in the region of the bonded pair would be drawn, via its valence electron, to the exposed core charge of the 2-atom cluster. It would contribute its valence electron to a bond with one or the other of the pair. The sodium atoms are therefore driven to aggregate, using their relatively few valence electrons to bond together as many nuclei as possible via the Coulomb attraction. The aggregation is very regular in nature, resulting in an ordered arrangement of metal atoms called a *lattice*. As a result of aggregation, each sodium atom uses all of its valence orbitals in forming partial bonds with a number of neighbours.

Contrast this with the covalent bonding between two chlorine atoms. The covalent bond is again localized, but the remaining valence electrons are distributed evenly around each chlorine atom, effectively shielding the core charge from chlorine atoms in other molecules. Further, all of the s and p valence orbitals are full. In order to aggregate as sodium does, chlorine would have to use either the 3d orbitals or the orbitals from the n = 4 main shell.

But the core charge experienced by an electron in either of these orbital sets is essentially zero (17-17 = 0). Consequently, chlorine is limited in its aggregation to the formation of covalently bonded diatomic molecules. As we will discuss shortly, there is only a very weak tendency for the diatomic molecules to aggregate via intermolecular forces. Interestingly, although hydrogen, like sodium, has only a single valence electron, it is not metallic. Instead, hydrogen atoms form covalently bonded diatomic molecules.

Why does hydrogen behave so differently from sodium and the other metals? Formation of a localized covalent bond between two hydrogen atoms results in exposure of the peripheral regions of the molecule to the core charge of the hydrogen atoms. In this case, because there are no core electrons, the H atom core charge is the nuclear charge of 1+. In the case of sodium and other metals, this situation leads to aggregation using the empty valence p orbitals. However, hydrogen has no valence p orbitals. The single orbital in the n=1 shell is full; no further aggregation is possible. The formation of metallic bonds can occur only if empty valence orbitals are available. Experiments have shown that if hydrogen is subjected to extremely high pressure, which forces the diatomic molecules to aggregate, it exhibits metallic characteristics.

In sodium, as in most metals, there are too few electrons to form normal two-electron bonds between all pairs of bonded atoms. To achieve stability, the valence electrons are spread over many nuclei, simultaneously bonding them all together. This spreading of electrons is called *delocalization*, and in some ways is similar to the delocalization that occurs in resonance. In metals, however, delocalization is much more extensive, and it is best to think of the electrons as waves that extend over the entire dimensions of the metallic crystal. A schematic structure of metallic bonding.

The nuclei are represented as points, and the electrons as waves. This picture enables us to understand the properties of metals. That the electrons are delocalized waves is consistent with the conductivity of metals, and with their flexibility and strength. Putting stress on metals moves the nuclei, but the waves adjust, maintaining bonding. Because the electrons are waves, they are everywhere in the crystal at once. They are mobile, rationalizing the electrical and thermal conductivity of metals.

Finally, the pooling of valence orbitals by aggregated sodium atoms leads to a large number of closely spaced orbitals in the metallic crystal. Thus the electrons absorb and reemit photons of all energies in the visible region of the spectrum, giving metals their highly reflective appearance. There are so few bonding electrons in metallic sodium that only about 1/8 of an electron pair bond holds each pair of adjacent sodium atoms together. It is little surprise, then, that sodium is very soft (weak bonding), very reactive (deficiency of bonding electrons), with low density, and low melting and boiling points.

Example: In magnesium, how many metallic bonds form per Mg atom?

Solution: The electron configuration of the magnesium atom is $1s^2\ 2s^2\ 2p^6\ 3s^2$. We assume that metallic bonds are formed using the valence electrons only. Because each Mg atom contributes 2 valence electrons to the metallic bonding, and because 2 electrons constitute one bond in Lewis terms, there is one bond per Mg atom in the metallic lattice.

Intermolecular Forces

Having described the forces that operate within molecules (including ionic crystals and metals), we are ready to discuss the interactions that occur between species (atoms, molecules, and ions) that are not covalently bonded to each other. These are called *intermolecular forces*; they can originate in a number of ways.

Dipole-Dipole Forces

Intermolecular forces that operate between neutral molecules having molecular dipole moments are called dipole-dipole forces. The structure of a molecule may cause its centers of positive and negative charge to be displaced from one another, giving the molecule a dipole moment. The dipole moments of two neighbouring molecules tend to align with the + end of one dipole near the - end of the other, so that forces of attraction between them are maximized.

The maximum force of attraction between two dipoles, m_1 and m_2, separated by a distance r is given by equation.

$$F = \mu_1 * \mu_2 / \Omega * r^4$$

Equation is Coulomb's Law, in disguised form. It is necessary to raise the distance of separation to the 4th power in the denominator to cancel the distance units that appear in both dipole moments in the numerator. Physically, the 4th power dependence means that the attraction between dipoles falls off much more rapidly with distance than would the attraction between isolated charges. This is because the + and - ends of dipole 2 are almost equidistant from dipole 1 when the dipoles are separated by a distance that is much greater than the dipole length.

The force of attraction between the + end of dipole 1 and the - end of dipole 2 is almost cancelled by the force of repulsion between the two + ends. Dipole-dipole forces operate between molecules of water, and between molecules of the substances.

Such forces are obviously much weaker than those operating in ionic or covalent network solids, and give rise to potential wells having depths in the approximate range 5-20 kJ/mole. Many molecular substances with dipolar molecules exist as liquids at ambient temperature, and have relatively low boiling points. In particular, many organic (carbon-containing) compounds are of this type.

Example: Which substances experience dipole-dipole intermolecular forces?

SiF_4,$CHCl_3$, CO_2, SO_2

Solution: We can draw the following conclusions about the molecules of these substances.

- SiF_4: Tetrahedral Si-F bonds are polar, but no molecular dipole; bond dipoles cancel
- CO_2: Linear C-O bonds are polar, but no molecular dipole; bond dipoles cancel
- SO_2: Bent S-O bonds are polar and do not cancel. Sulfur lone pair dipole only partially offsets net bond dipole.
- $CHCl_3$: Tetrahedral C-H and C-Cl bonds are polar and do not cancel
- SO_2 and CHCl3 experience dipole-dipole intermolecular forces.

Hydrogen Bonding

Hydrogen bonding (H-bonding) is a special kind of dipole-dipole force that occurs when a hydrogen atom is bonded to one of the very electronegative atoms, F, O, or N. The H-F, H-O, and H-N bonds are very polar, because the electronegative atom draws the bonding electron pair strongly to itself.

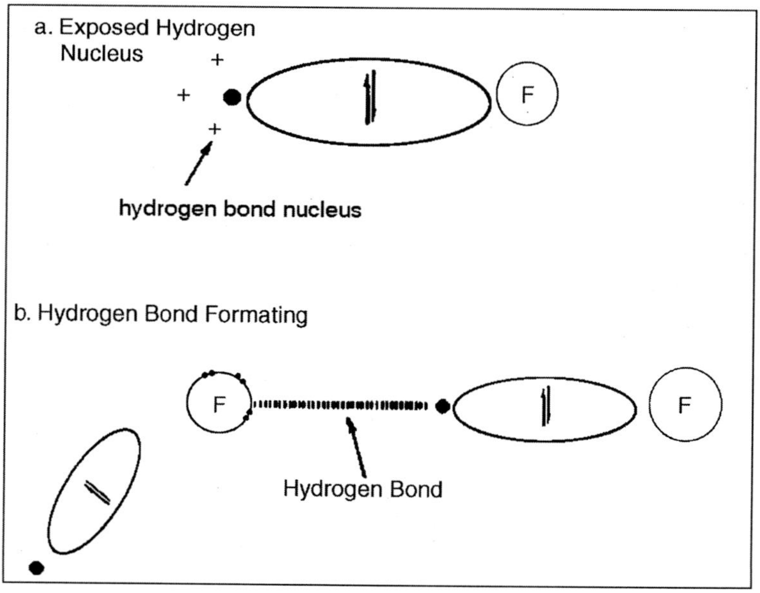

Figure: *Hydrogen Bonding*

This leaves the hydrogen nucleus exposed. The resulting bond dipole is substantial. The positive end of this dipole, consisting essentially of an exposed hydrogen nucleus, has a high charge density because the proton is so small. Consequently, it strongly attracts the negative end of a similar bond dipole in another molecule to give a structure.

The hydrogen atom is sandwiched between the two electronegative atoms in a linear arrangement. Hydrogen bond forces cause potential wells of depth in the range 5-50 kJ/mole. Note that these are on average deeper than those resulting from ordinary dipole-dipole forces. The intermolecular forces between water molecules are of the hydrogen bonding type. They are responsible for the abnormally high boiling point of water, which, with its small molecular weight, would otherwise be a gas at room temperature. Compare the boiling point of water (100 °C) with that of methane (-162 °C), which has a similar molecular weight, but lacks hydrogen bonding and ordinary dipole-dipole forces. Hydrogen bonding is responsible for the formation of genes in the DNA molecule; for the helical structure of proteins; and for the incredible strength of Kevlar, a DuPont polymer used for canoe hulls and bullet-proof vests.

Example: In which substances do hydrogen bonding forces operate between molecules?

$$CHCl_3, CH_3CH_2OH, HNO_3, PH_3$$

Solution: Hydrogen is bonded to one of the very electronegative atoms in CH_3CH_2OH and HNO_3. Hydrogen bonding should occur in both of these substances.

Instantaneous Dipole-Induced Dipole (Dispersion) Forces

It may seem that there would be no forces of attraction between atoms of helium, because they are uncharged and have no permanent dipole moment. Yet He can be liquified at the very low temperature of 4 K. That this is possible indicates that intermolecular forces do exist, though they are very weak. What is their origin? We portray the He atom as if the electron distribution about the nucleus is perfectly spherical. However, the electrons are constantly in motion and it is possible at any given instant for the electron cloud to be skewed slightly to one side of the nucleus.

This generates an instantaneous dipole moment in the He atom. During its transitory existence, this dipole induces a dipole in a neighbouring atom and the two instantaneous dipoles attract. Forces between transitory dipoles are called instantaneous dipole-induced dipole forces, or alternately, London dispersion forces after the scientist who first proposed them. We will refer to them as *dispersion forces.*

It is these forces that are responsible for the liquefaction and/or solidification of a number of substances whose molecules do not possess permanent dipole moments. For example, dry ice is solid CO_2. The CO_2 molecule is linear and, though it has non-zero bond dipoles, has no overall dipole moment.

Similarly, CCl_4, with perfectly tetrahedral molecules and no net dipole moment, is a liquid at room temperature and up to 78 °C. And iodine, which consist of I_2 molecules, is a solid at room temperature. Dispersion forces depend on two features of molecular structure. First, they increase in magnitude with the size and distortability (usually called the polarizability) of the electron clouds of the interacting particles. Size and polarizability increase as molecular weight increases.

It follows that dispersion forces increase as molar mass increases. For substances of large atomic or molecular mass, dispersion forces are strong enough that the substances are solid or liquid at room temperature. We saw this above for CCl_4 and I_2. A striking example is provided by the diatomic halogens of group 17, which progress from F_2, a gas at room temperature and 1 atm pressure, to Cl_2, also a gas, to Br_2, a liquid, and finally to solid I_2. Only dispersion forces are operative in this series of substances. Second, dispersion forces depend upon molecular shape via the surface area over which two molecules can be in contact.

The larger the surface area of contact, the stronger the dispersion forces. Molecules that are roughly spherical in shape are able to contact each other only minimally. In contrast, molecules that are planar or linear in shape can maintain a large surface area of contact, with correspondingly larger dispersion forces. The classic example of the effect of molecular symmetry (shape) on the magnitude of dispersion forces is provided by the series of isomeric pentanes (C_5H_{12}).

Neopentane is highly symmetrical and nearly spherical in shape. Intermolecular forces are relatively small and the boiling point is correspondingly low. Pentane is linear, so two pentane molecules are able to contact each other along the entirety of their length. Intermolecular forces are relatively large, and the boiling point is correspondingly high. Isopentane is intermediate structurally, and the least symmetrical of the three isomers. Intermolecular forces are less than in pentane, but greater than in neopentane. The boiling point is intermediate. Graphite, the most stable form of elemental carbon at ordinary temperature and pressure, has the structure. Each carbon atom is bonded to three others in trigonal planar fashion, using sp^2 hybrid orbitals.

The unhybridized p orbitals are used to form a network of pi bonds that extends over the entire array of bonded carbons, producing a large planar sheet of carbon atoms. These sheets are then stacked one atop the other to produce the three dimensional structure of graphite. Interestingly, there are no covalent bonds holding one sheet to the next. The sheets attract one another via dispersion forces, which are very strong because of the large surface area of contact.

It is important to realise that dispersion forces operate between all molecules, whether or not other forces also operate. Molecules of chloroform, $CHCl_3$, are atttracted by a combination of dipole-dipole and dispersion forces. The magnitude of dispersion forces varies with molar mass; however, they are generally weaker than dipole-dipole forces, giving potential wells in the range 0.1 to 5 kJ/mole.

Example: Arrange the following non-polar molecules in order of increasing melting point.

$$SiF_4,\ CS_2,\ CI_4,\ GeCl_4$$

Solution: None of these molecules is expected to have a dipole moment. Only dispersion forces will operate, and these increase with increasing molecular mass. The molar masses of these substances follow:

Substance	***MM***
SiF_4	104.077
CS_2	76.131
CI_4	519.631
$GeCl_4$	214.402

The intermolecular forces, and the melting points, should increase in the following order:

$$CS_2 < SiF_4 < GeCl_4 < CI_4$$

The experimentally determined melting points are –110.8, –90, –49.5, and 171 °C, respectively.

Example: Molar masses, dipole moments, and boiling points for several covalently-bonded substances are given below. Rationalize the observed boiling points in terms of intermolecular forces.

Substance	*MM*	*m*	*Boiling Point, K*
He	4.003	0	4
H_2	2.016	0	20
N_2	28.014	0	77
CO	28.010	0.33	83
HCl	36.461	3.60	188
HBr	80.912	2.67	206
SO_2	64.058	5.42	263
HF	20.006	6.37	292
CCl_4	153.823	0	350
H_2O	18.015	6.17	373

Solution: The substances fall into two categories, those with dipole moments and those without. Only dispersion forces operate in substances without dipole moments. Selecting from the table those substances with no dipole moment, shown below, we see that, with the exception of hydrogen, boiling point increases with increasing MM, as expected from our discussion of dispersion forces above. The dipole moment of CO is so small that it seems reasonable to include it in this list. Its boiling point is only slightly larger than that of N_2, due to weak dipole-dipole forces in addition to the dispersion forces.

Substance	*MM*	*m*	*Boiling Point, K*
He	4.003	0	4
H_2	2.016	0	20
N_2	28.014	0	77
CCl_4	153.823	0	350

Substances having substantial dipole moments are listed below, now in order of increasing dipole moment.

Substance	*MM*	*m*	*Boiling Point, K*
HBr	80.912	2.67	206
HCl	36.461	3.60	188
SO_2	64.058	5.42	263
H_2O	18.015	6.17	373
HF	20.006	6.37	292

We see only a rough correlation of boiling point with dipole moment: HBr and HCl have boiling points on the low end of the scale; SO_2 is intermediate; and H_2O and HF have high-end boiling points.

However, the boiling points of HBr and HCl are inverted based on the size of the dipole moment. HCl, with the larger dipole moment, has the lowest boiling point in the group of five. Here is clear evidence that dispersion forces operate in addition to dipole-dipole forces. Dispersion forces are substantially larger for HBr (MM 80.9) than for HCl (MM 36.5). This is sufficient to raise the boiling point for HBr above that for HCl.

The remaining 3 substances do not yield to the same argument, however. The MM of SO_2 is more than three times larger than those of HF and H_2O, and its dipole moment is only somewhat less than that for water. Yet the boiling points of HF and particularly of H_2O are strikingly higher than that for SO_2. Here we have evidence of the potency of hydrogen bonding intermolecular forces, which operate in both HF and H_2O. Despite its small MM, HF boils 30 ° higher than SO_2 because hydrogen bonding causes the potential well of the liquid to be relatively deep. The boiling point for water is about 80 ° higher than that for HF because each water molecule can engage in up to four hydrogen bonds (one for each H atom, and one for each lone pair on the O atom).

Mixed Forces in Solutions

All of the forces that can operate at the molecular level within a single pure substance are discussed above. It is possible to have what we might call hydrid forces that operate when pure substances are mixed to form solutions. When an ionic solid such as sodium chloride dissolves in water, ion-dipole forces are responsible. The maximum magnitude of ion-dipole forces is given by a form of Coulomb's Law that is intermediate between equations.

$$F = q^*\mu/\Omega^*r^3$$

Here q is the charge on the ion, m is the dipole moment of the solvent, and r is the distance between the ion and the center of the solvent dipole.

Example: Consider the following three situations:

- A charge of 1+ separated by 2.0×10^{-10} m from a charge of 1– (this might be two ions in a crystal);
- A charge of 1+ separated by 2.0×10^{-10} m from a dipole of magnitude 1.6×10^{-29} C-m (a dipole of this magnitude would result from charges of 1+ and 1– separated by a distance of 1.0×10^{-10} m).
- Two dipoles, each of magnitude 1.6×10^{-29} C-m, separated by 2.0×10^{-10} m.

Compare the magnitudes of the attractive forces operating in these situations. This will provide an idea of the relative magnitudes of ion-ion, ion-dipole, and dipole-dipole forces. Then, compare the forces when the distance of separation of the interacting entities is increased by a factor of 4 to 8.0×10^{-10} m.

Solution: We apply equations:

Ion-ion: $F = (1.6 \times 10^{-19}\ C)^2/\Omega^*(2.0 \times 10^{-10})^2 = 6.4 \times 10^{-19}/\Omega$ Newtons

Ion-dipole $F = (1.6 \times 10^{-19})(1.6 \times 10^{-19})/\Omega^*(2.0 \times 10^{-10})^3 = 3.2 \times 10^{-19}/\Omega$ Newtons

Dipole-dipole $F = (1.6 \times 10^{-29})^2/\Omega^*(2.0 \times 10^{-10})^4 = 1.6 \times 10^{-19}/\Omega$ Newtons

The ion-ion force is stronger by a factor of 4 than the dipole-dipole force. The ion-dipole force is intermediate. When the separation distance is increased by a factor of 4, the magnitudes of all forces decrease because they depend inversely on a power of the distance.

The ion-ion force is reduced by a factor of 4^2; the ion-dipole force by a factor of 4^3; and the dipole-dipole force by a factor of 4^4:

Ion-ion: $F = 4 \times 10^{-20}/\Omega$ NtIon-dipole $F = 0.5 \times 10^{-20}/\Omega$ NtDipole-dipole $F = 0.063 \times 10^{-20}/\Omega$ Nt

The forces involving dipoles clearly decrease much more rapidly with increasing distance of separation than do ion-ion forces. Whether or not two liquids are mutually soluble (miscible) depends on whether the mixed intermolecular forces can compete with the forces operating within each pure liquid alone.

The intermolecular forces in pure water are primarily hydrogen bonding (dipole-dipole) forces, superimposed on weak dispersion forces.

Those in pure acetone are dipole-dipole forces superimposed on dispersion forces that are stronger than the dispersion forces in water. When acetone and water are mixed, they readily dissolve in one another in all proportions, apparently because the dipole-dipole forces between water and acetone molecules are of strengths comparable to those operating in the two separate liquids.

On the other hand, only dispersion forces operate within pure carbon tetrachloride, because the CCl_4 molecule has a molecular dipole moment of zero. Because the molecules are spherical, giving small surface area of contact, these forces are fairly weak. Consequently, carbon tetrachloride and water are not mutually soluble. They are said to be *immiscible*. When CCl_4 and water are mixed, the two liquids segregate into layers, with the denser liquid, chloroform, on the bottom. The liquid-liquid interface (boundary) is clearly visible.

Finally, a non-polar solid like I_2 dissolves readily in a non-polar liquid like CCl_4, because the mixed dispersion forces are comparable in magnitude to the dispersion forces in pure CCl_4 and pure I_2.

However, the solubility of I_2 in water is essentially zero, because mixed forces cannot compete with the strong dipole-dipole forces in water.

Water molecules are unwilling to separate in order to make room for the non-polar I_2 molecules. In this way, we can extend our intermolecular force ideas to provide an understanding of the formation of mixtures.

Summary of Mixed-Force Concepts

- Substance A will dissolve in substance B if the forces operating between A and B molecules are of similar type (dipolar or dispersion) and strength to those operating between A molecules and between B molecules. This statement is often sloganized: "Like dissolves like."
- Substance A will not dissolve in substance B if the forces operating between A and B molecules are of different types (one dipolar, one dispersion) (and therefore usually different strengths) than those operating between A molecules and between B molecules. This can also be sloganized: "Like does not dissolve unlike."

Example: For each pair, predict whether substance A will dissolve in liquid B, and explain your reasoning.

Substance A	***Liquid B***
C_2H_5OH	CH_3OH
Br_2(l)	C_5H_{12}
Br_2(l)	H_2O
C_2H_5OH	H_2O
C_2H_5OH	$CHCl_3$
NaCl	$CHCl_3$
NaCl	CH_3CH_2OH
H_2O	CH_2Cl_2

Solution: C_2H_5OH (ethanol) and CH_3OH (methanol) are both liquids due to fairly strong H-bonding intermolecular forces. Ethanol is expected to dissolve in methanol. Br_2(l) and C_5H_{12} (pentane) are both non-polar liquids with dispersion forces operating between molecules. Thus they are "like" and we expect bromine to dissolve in pentane.

Br_2 and H_2O are "unlike." Bromine will not dissolve readily in water. Both liquids exert H-bonding forces at the molecular level. Ethanol is expected to dissolve in water.

This is not obvious. At first glance, we would call ethanol and chloroform "unlike" because H-bonding forces operate in ethanol, whereas dispersion forces operate in chloroform.

However, in addition to its polar -OH group, ethanol has a non-polar hydrocarbon portion that is capable of exerting dispersion forces on molecules of chloroform. These two liquids are therefore miscible. Ionic NaCl will certainly not dissolve in non-polar chloroform.

Because ethanol is polar at the -OH end, we might expect NaCl to dissolve to some extent in ethanol, but not to the same extent as in water. This is indeed found. Polar water will not mix with non-polar dichloromethane.

Example: Describe what the system will look like during and following each process:

- 50 mL of distilled water is added to an Erlenmeyer flask.
- 0.10 g of I_2(s) is added to the flask, and the flask is stoppered and vigorously shaken.
- 50 mL of CCl_4 is added to the Erlenmeyer flask, and the flask is stoppered and shaken.

Solution

- The clear, colourless liquid will occupy the bottom of the flask.
- I_2, and intensely violet solid, is not expected to be soluble in water because of incompatible intermolecular forces. Because I_2 is more dense than water, it will sink to the bottom of the flask. The water will remain clear and colourless.
- CCl_4 is a non-polar liquid in which the intermolecular forces are of the dispersion type. It is expected to be immiscible with water. Because it is more dense than water, it will sink to the bottom of the flask, and the water will float on it with a distinct liquid-liquid boundary. CCl_4 and I_2 have compatible intermolecular forces, so I_2 is expected to dissolve readily

in carbon tetrachloride to give a purple solution. The flask will thus contain a layer of purple liquid under a layer of colourless liquid.

Distribution of a Solute Between Two Immiscible Liquids in Contact

When two immiscible liquids, such as water and chloroform ($CHCl_3$), are placed in contact, they will mutually exclude each other, forming two liquid layers. The less dense liquid will float on the denser one, and a boundary surface will be clearly visible. It is interesting to think about what will happen if a third substance, solid or liquid, is added to the system consisting of two immiscible liquids, A and B.

If the third substance, C, is soluble in one of the liquids (say A) and insoluble in the other (B), it will dissolve exclusively in the A layer, as in example 6-14. More interesting is the situation in which the C is soluble in both liquids to at least some extent.

In this case, we find that the substance C *partitions*, or *distributes*, itself between the two liquid layers. We further find that the concentration of the third substance is higher in the liquid with which its intermolecular forces are most compatible.

The ratio of the concentration of C in liquid A to that in liquid B is found to attain a constant value K_d, defined as in equation:

$$K_d = [C]_A/[C]_B$$

This constant ratio is established no matter how the 3-component mixture is prepared.

Three such ways follow:

1. A and B are first brought into contact. C is then added, and the container is shaken.
2. C is first dissolved in A. A (containing C) and B are then brought into contact, and the container is shaken.
3. C is first dissolved in B. A and B (containing C) are then brought into contact, and the container is shaken.

In addition, the ratio is reestablished if the mixture is altered in any of the following ways:

- More A is added.
- More B is added.
- More C is added.

We can understand this behaviour in terms of a simple layers of A and B in contact at a boundary, and molecules of C initially dissolved in liquid A.

The boundary between the two liquids provides a "doorway" from one to the other through which C molecules, which have reasonable interactions with both A and B molecules, can pass.

Thus molecules of C will tend to move across the boundary from liquid A, where they are initially dissolved, into liquid B, in which there are initially no molecules of C. In other words, they begin to distribute themselves between the two liquids. After some time, enough C molecules will have passed into liquid B that the rates of passage of C from A to B and back will be the same.

Once this situation is attained, there will be no further change in the concentration of C in the two layers. In other words, the constant ratio in equation will be attained.

From the magnitude of K_d, we may easily deduce which liquid C prefers. If $K_d < 1$, C prefers B; if > 1, C prefers A. Distribution of a substance between two liquid phases is an example of *dynamic equilibrium*, a situation in which two opposing processes (here, movement of C both ways across the boundary) occur at equal rates. We will have much more to say about dynamic equilibrium later.

A Nod to Other Classification Systems

Classification of the various types of forces as "inter-" or "intramolecular" is not absolute.

Thus you may see the forces classified as follows:

- Covalent Bonds (in molecular and network covalent substances)
- Ionic Bonds
- Metallic Bonds
- Interunit forces
  - Ion-dipole
  - Dipole-dipole
  - Hydrogen bonds
  - Dispersion forces

In this classification scheme, ionic and metallic bonds are categorized separately rather than as intermolecular forces, because there are no discreet molecules in ionic compounds or metals. Another classification system has recently become popular and is experiencing increasing use.

This goes as follows:

- Covalent Interactions (includes covalent bonds in molecular and network covalent substances)
- Non-covalent Interactions
  - Ionic bonds
  - Metallic bonds
  - Metal-ligand interactions
  - ion-dipole forces
  - dipole-dipole forces
  - hydrogen bonds
  - dispersion forces

Here there are only 2 major categories: covalent, and not covalent! It is important to realise that the distinction between covalent bonding and some of the non-covalent interactions is blurred.

For example, many people believe that metal-ligand interactions and even hydrogen bonding are weak covalent interactions. The bottom line is that it is not particularly important how the classification is done, as long as the various types of interactions are understood. Pick your own favourite classification system, and use it!

The Interplay Between Molecular Kinetic and Potential Energies

The concept of temperature is of profound significance in science. We take it for granted in our daily lives, without thinking much about what it is. Temperature is themanifestation of molecular kinetic energy. The steam rising from a pan of boiling water feels hot because the molecules of water vapour are moving rapidly. Their collisions with the skin transfer kinetic energy, which we interpret as heat. The relationship between temperature and the average translational kinetic energy of an atomic, ionic, or molecular unit is very simple.

$$\text{Average KE}_{\text{molecule}} = 3/2\ kT$$

T is the absolute (Kelvin) temperature, and k is a fundamental constant of nature called Boltzmann's constant. Its value is 1.381 ×

10^{-23} J/K. By $KE_{molecule}$, we signify the kinetic energy of any microscopic unit of matter. This can be anything from an electron to a DNA molecule.

Keep in mind in using equation that the kinetic energy referred to is the translational kinetic energy—the energy of motion of the particle as a whole through space. Translational kinetic energy is often called thermal energy.

Example: Calculate the thermal energy of one mole of nitrogen gas at room temperature, 293 K.

Solution: Calculate the kinetic energy of an average molecule; then multiply by Avogadro's Number.

$$\text{Average } KE_{molecule} = 3kT/2 = 3(1.381 \times 10^{-23}\ \text{J/K})(293\ \text{K})/2$$

$$= 6.07 \times 10^{-21}\ \text{J/molecule}$$

$$KE_{mole} = N_o KE_{molecule} = 3N_o kT/2$$

$$= 6.07 \times 10^{-21}\ \text{J/molecule} \times 6.023 \times 10^{23}\ \text{molecules/mole}$$

$$= 3.66 \times 10^{3}\ \text{J/mole}$$

The product of Boltzmann's constant and N_o is a constant, R, which occurs in the ideal gas law. Its value is 8.314 J/K-mole. Expressed in terms of R, the kinetic energy per mole of gas is 3RT/2. We have seen that molecular forces promote aggregation by lowering molecular potential energy.

Kinetic energy counteracts the effects of intermolecular forces. These ideas lead to a very useful generalization: Potential energy favours aggregation; kinetic energy opposes aggregation. The stronger the molecular forces, the more rapidly atoms, molecules, or ions must move in order to overcome them.

Even the strongest forces, however, may be overcome by the vigour of molecular motion if the temperature is high enough. At a temperature that gives a kinetic energy comparable in magnitude to the depth of the potential well, molecules begin to escape the potential well, and the substance undergoes a phase change (that is, sublimes, melts, or boils).

A phase change may therefore by viewed as an exchange of KE for PE. If the potential well is shallow, as in liquid helium (where the only forces are the very weak dispersion type) the required kinetic energy is achieved at very low temperature (4 K). If the potential well is deep, as in the ionic compound MgO (where the forces are the strong ion-ion type), very high temperature is required for particles to escape (3125 K).

Table shows the parallel between the potential well depth and the temperature of melting for a number of substances.

Table: *Melting Temperature and Potential Well Depth*

Substance	***Potential Well Depth, J/mole***	***Melting T, K***
He	0.105	3.5
H_2	1.04	14.0
Xe	14.7	161
N_2	6.30	63.2
Cl_2	26.8	172.1
I_2	56.3	387
H_2O	46.7	273
CO_2	33.6	217
NaCl	771	1074

Try plotting melting point against potential well depth to see how good the correlation is.

Consequences of Intermolecular Forces

The operation of the types of forces between molecules, atoms, and ions has profound implications for the properties, function, and reactivity of chemical substances. By way of illustration, we discuss intermolecular forces in that most fundamental and essential of all molecules, water. The water molecule is bent, with a bond angle of about 104 °. This shape is understandable in terms of the VSEPR theory. A consequence of this shape is that the molecule is polar (*i.e.*, has a molecular dipole).

The molecular dipole moment and the fact that water contains hydrogen bonded to the very electronegative oxygen atom leads to strong intermolecular forces of the hydrogen bonding (dipole-dipole) type. A consequence of the existence of these forces is that the melting and boiling points of water are much higher than would be expected on the basis of molar mass (*e.g.*, the boiling point of methane, with molar mass 16, is –162 °C). Thus water is a liquid or solid under temperature conditions found on the earth's surface.

As a liquid, it provides a medium for the genesis and sustenance of life. Another consequence of polarity is that water has excellent solvating properties: it dissolves substances as diverse in structure as salts, sugars, and huge protein molecules that function as enzymes.

Due to the shape of the water molecule and the intermolecular hydrogen bonding, water adopts an open lattice structure when it freezes, in which each oxygen atom is covalently bonded to two hydrogen atoms and hydrogen bonded to two more via its two lone pairs of electrons. A consequence of his open lattice is that ice is less dense than liquid water, and floats.

Lakes therefore freeze top down, the layer of ice on top serving to insulate the unfrozen liquid below and thereby to protect aquatic life. If, like most substances, water were more dense as a solid, ice would sink, entire lakes would freeze, and life as we know it would not have arisen. (Of course, the density inversion of water has negative consequences, to, with which we are all familiar: potholes in roads, erosion of rock formations, and the like.)

Finally, the structure of the water molecule and the resulting intermolecular interactions are at the root of the ability of water to function as both an acid and a base. This aspect of the reactivity of water is crucial in many biochemical and geological processes.

Chapter

3

Biogeochemical Cycle

An ecosystem is a group of living and non-living components interacting together on a given physical landscape. The size of an ecosystem is arbitrary and could be as small as a few square centimeters if you are looking at a soil microbial ecosystem; as large as thousands of square kilometers if you are looking a biome like the Great Plains ecosystem; or a few hectares if you are looking at a single forest stand ecosystem.

Forested Ecosystem

An ecosystem model is an accurate but simplified representation of an ecosystem that can be very useful in thinking about or simulating the actions of a real ecosystem. Because any ecosystem has many different but interrelated components, the best way to understand the system is to break it down into its component parts. To get an introduction to a very simplified forest model, which gives participants and introduction to how a hardwood forest ecosystem works before and after exotic earthworms invade.

- *Step One*: The first step in building a graphical model of a hardwood forest ecosystem is to identify its major components. The components of any ecosystem are those physical things that contain energy and nutrients. A forested ecosystem, by definition contains trees, so that is our first component. In addition to the various species and layers of trees in a forest, there are other distinct ecosystem components. The *understory* contains most of the visible plant life found between the sapling layer and forest floor. We will add those two components to our forest ecosystem model later

- *Step Two*: Once you have identified the components of your ecosystem model, you need to define the *processes* that connect the components. This is graphically done by using arrows to indicate the flow of nutrients or energy among the different ecosystem components. The components of our ecosystem model are now connected by processes that result in the movement of energy or nutrients among the components.

  One thing to notice in our ecosystem model is that there is no process connecting the *trees* component with the understory plants component. This is because there are no substantial processes that result in the flow of nutrients directly from a tree to an understory plant or visa-versa. There would be important relationships between the trees and understory plants. Trees provide shade to the understory plants. But remember, in an ecosystem model, only processes that result in flow of energy or nutrients are represented. Now let's add the animals and the people components to our ecosystem. energy and nutrients flow from the trees and understory plants to the animals when they eat the leaves, twigs and buds of trees or graze on understory plants; and when the animal excrete waste products or die, energy and nutrients are returned to the forest floor component.

- *Step Three*: Determine the major inputs and outputs of your ecosystem. As you are building your ecosystem model, one thing to think about is whether your ecosystems could be opened or closed. A closed ecosystem is one that has no inputs of energy or nutrients from outside the ecosystem and no outputs of energy or nutrients leaving the system. The earth is an example of a closed ecosystem with respect to nutrients and an open ecosystem with respect to energy All the nutrients that have ever been on earth are here and simply continue to cycle, there are no additions or losses. However, the earth is constantly getting inputs of energy from the sun and simultaneously radiating energy back.

  The earth ecosystem and has no inputs or outputs of nutrients which are constantly recycled within the global ecosystem, while the earth has both inputs and outputs of energy that are in a relatively stable equilibrium.

Just as the Earth ecosystem is closed with respect to nutrients, unmanaged earthworm-free hardwood forest ecosystems are often very nearly closed nutrient ecosystems that there are very few inputs or outputs of nutrients. Rather the nutrients are constantly recycled among the various ecosystem components. In contrast, most agricultural ecosystems require nutrient inputs from outside to function properly some typical inputs and outputs of nutrients and energy for forested ecosystems include evapotranspiration, nutrient leaching, sunlight and rain.

- *Step Four*: Once you have identified the components, processes and major inputs and outputs in your ecosystem model, then you can begin to add the actual values to these parts of your ecosystem by measuring them. The amount of litter that falls to the forest floor each year (a process), what the biomass of trees is in a given forest (a component), how much light reaches the forest over a growing season (an input), or how much nitrogen leaches from the forest (an output). In the case of leaf litter, you can put out trays in the forest and after all the leaves have fallen for the year, dry and weight the leaf liter in your trays. They you can use that value to calculate an estimate of the total leaf litter for your forest
- *Step Five*: Use your ecosystem model to think about how changes can cascade through an ecosystem or to ask specific questions that can be answered with further research. When the major components, processes and inputs and outputs of an ecosystem are understood, then you can use the model how changing one part of the ecosystem affects other parts. If you harvest trees from your forest, that will decrease the amount of leaf litter reaching the forest floor each year which may lead to decreases in available nutrients for understory plants.

An ecosystem that is in equilibrium doesn't gain or lose nutrients. Researchers may monitor soil nutrient levels for many years after trees have been harvested how the real forest behaves compared to what they thought might happen based on their forest model, their understanding of how the forest works.

If the results in the real forest are very different than those predicted by their model, then they know that they don't have full understanding of how their forest works and they may go back to try to improve their model.

In general, when building a model, you start simple and add more detail as needed. To get an introduction to a very simplified forest model which gives participants and introduction to how a hardwood forest ecosystem works before and after exotic earthworms invade.

Energy Flow Through the Ecosystem

Both energy and inorganic nutrients flow through the ecosystem. We need to define some terminology first. Energy "flows" through the ecosystem in the form of carbon-carbon bonds. When respiration occurs, the carbon-carbon bonds are broken and the carbon is combined with oxygen to form carbon dioxide. This process releases the energy, which is either used by the organism or the energy may be lost as heat. The dark arrows represent the movement of this energy.

The other component are the inorganic nutrients. They are inorganic because they do not contain carbon-carbon bonds. These inorganic nutrients include the phosphorous in your teeth, bones, and cellular membranes; the nitrogen in your amino acids; and the iron in your blood.

The movement of the inorganic nutrients is represented by the open arrows. Note that the autotrophs obtain these inorganic nutrients from the inorganic nutrient pool, which is usually the soil or water surrounding the plants or algae.

These inorganic nutrients are passed from organism to organism as one organism is consumed by another. Ultimately, all organisms die and become detritus, food for the decomposers. At this stage, the last of the energy is extracted and the inorganic nutrients are returned to the soil or water to be taken up again. The inorganic nutrients are recycled, the energy is not.

Many of us, when we hear the word "nutrient" immediately think of calories and the carbon-carbon bonds that hold the caloric energy. When writing about energy flow and inorganic nutrient flow in an ecosystem, you must be clear as to what you are referring. Unmodified by "inorganic" or "organic", the word "nutrient" can leave your reader unsure of what you mean.

This is one case in which the scientific meaning of a word is very dependent on its context. Another example would be the word "respiration", which to the layperson usually refers to "breathing", but which means "the extraction of energy from carbon-carbon bonds at the cellular level" to most scientists (except those scientists studying breathing, who use respiration in the lay sense).

To summarize: In the flow of energy and inorganic nutrients through the ecosystem, a few generalizations can be made:

- The ultimate source of energy (for most ecosystems) is the sun
- The ultimate fate of energy in ecosystems is for it to be lost as heat.
- Energy and nutrients are passed from organism to organism through the food chain as one organism eats another.
- Decomposers remove the last energy from the remains of organisms.
- Inorganic nutrients are cycled, energy is not.

Food Chains and Webs

A food chain is the path of food from a given final consumer back to a producer. For instance, a typical food chain in a field ecosystem might be:

Grass → Grasshopper → Mouse → Snake → Hawk

Note that even though I said the food chain is the path of food from a given final consumer back to a producer we typically list a food chain from producer on the left to final consumer on the right. Note to international readers: In Hebrew or Aramaic, or other languages which are read right-to-left, is it customary to list the food chains in the reverse order? By the way, you should be able to look at the food chain above and identify the autotrophs and heterotrophs, and classify each as a herbivore, carnivore, etc. You should also be able to determine that the hawk is a quaternary consumer.

The real world, of course, is more complicated than a simple food chain. While many organisms do specialize in their diets (anteaters come to mind as a specialist), other organisms do not. Hawks don't limit their diets to snakes, snakes eat things other than mice, mice eat grass as well as grasshoppers, and so on.

A more realistic depiction of who eats whom is called a food web. It is when we have a picture of a food web in front of us that the definition of food chain makes more sense. We can now a food web consists of interlocking food chains, and that the only way to untangle the chains is to trace *back* along a given food chain to its source.

The food webs are *grazing food chains* since at their base are producers which the herbivores then graze on. While grazing food chains are important, in nature they are outnumbered by *detritus-based food chains*. In detritus-based food chains, decomposers are at the base of the food chain, and sustain the carnivores which feed on them. In terms of the weight (or biomass) of animals in many ecosystems, more of their body mass can be traced back to detritus than to lving producers.

Pyramids

The concept of biomass is important. It is a general principle that the further removed a trophic level is from its source (detritus or producer), the less biomass it will contain (biomass here would refer to the combined weight of all the organisms in the trophic level).

This reduction in biomass occurs for several reasons:

- Not everything in the lower levels gets eaten
- Not everything that is eaten is digested
- Energy is always being lost as heat

It is important to remember that the decrease in number is best detected in terms or biomass. Numbers of organisms are unreliable in this case because of the great variation in the biomass of *individual* organisms. For instance, squirrels feed on acorns. The oak trees in a forest will always outnumber the squirrels in terms of combined weight, but there may actually be more squirrels than oak trees.

Remember that an individual oak tree is huge, weighing thousands of kilograms, while an individual squirrel weighs perhaps 1 kilogram at best. There are few exceptions to the pyramid of biomass scheme. One occurs in aquatic systems where the algae may be both outnumbered and outweighed by the organisms that feed on the algae.

The algae can support the greater biomass of the next trophic level only because they can reproduce as fast as they are eaten. In this way, they are never completely consumed.

It is interesting to note that this exception to the rule of the pyramid of biomass also is a partial exception to at least 2 of the 3 reasons for the pyramid of biomass. While not all the algae are consumed, a greater proportion of them are, and while not completely digestible, algae are far more nutritious overall than the average woody plant is (most organisms cannot digest wood and extract energy from it).

Biological Magnification

Biological magnification is the tendency of pollutants to become concentrated in successive trophic levels. Often, this is to the detriment of the organisms in which these materials concentrate, since the pollutants are often toxic. Biomagn-ification occurs when organisms at the bottom of the food chain concentrate the material above its concentration in the surrounding soil or water. Producers, as we saw earlier, take in inorganic nutrients from their surroundings. Since a lack of these nutrients can limit the growth of the producer, producers will go to great lengths to obtain the nutrients.

They will spend considerable energy to pump them into their bodies. They will even take up more than they need immediately and store it, since they can't be "sure" of when the nutrient will be available again (of course, plants don't think about such things, but, as it turns out, those plants, which, for whatever reason, tended to concentrate inorganic nutrients have done better over the years).

The problem comes up when a pollutant, such as DDT or mercury, is present in the environment. Chemically, these pollutants resemble essential inorganic nutrients and are brought into the producer's body and stored "by mistake". This is the first step in biomagnification; the pollutant is at a higher concentration inside the producer than it is in the environment.

The second stage of biomagnification occurs when the producer is eaten. Remember from our discussion of a pyramid of biomass that relatively little energy is available from one trophic level to the next. This means that a consumer (of any level) has to consume a lot of biomass from the lower trophic level. If that biomass contains the pollutant, the pollutant will be taken up in large quantities by the consumer. Pollutants that biomagnify have another characteristic. Not only are they taken up by the producers, but they are absorbed and stored in the bodies of the consumers.

This often occurs with pollutants soluble in fat such as DDT or PCB's. These materials are digested from the producer and move into the fat of the consumer. If the consumer is caught and eaten, its fat is digested and the pollutant moves to the fat of the new consumer. In this way, the pollutant builds up in the fatty tissues of the consumers. Water-soluble pollutants usually cannot biomagnify in this way because they would dissolve in the bodily fluids of the consumer.

Since every organism loses water to the environment, as the water is lost the pollutant would leave as well. Alas, fat simply does not leave the body. The "best" example of biomagnification comes from DDT. This long-lived pesticide (insecticide) has improved human health in many countries by killing insects such as mosquitoes that spread disease. On the other hand, DDT is effective in part because it does not break down in the environment. It is picked up by organisms in the environment and incorporated into fat. Even here, it does no real damage in many organisms (including humans). In others, however, DDT is deadly or may have more insidious, long-term effects. In birds, for instance, DDT interferes with the deposition of calcium in the shells of the bird's eggs.

The eggs laid are very soft and easily broken; birds so afflicted are rarely able to raise young and this causes a decline in their numbers. This was so apparent in the early 1960's that it led the scientist Rachel Carson to postulate a "silent spring" without the sound of bird calls. The banning of DDT, the search for pesticides that would not biomagnify, and the birth of the "modern" environmental movement in the 1960's. Birds such as the bald eagle have made comebacks in response to the banning of DDT in the US. Ironically, many of the pesticides which replaced DDT are more dangerous to humans, and, without DDT, disease (primarily in the tropics) claims more human lives.

Human vs. Natural Food Chains

Human civilization is dependent on agriculture. Only with agriculture can a few people feed the rest of the population; the part of the population freed from raising food can then go on to do all the things we associate with civilization. Agriculture means manipulating the environment to favour plant species that we can eat. In essence, humans manipulate competition, allowing favoured species (crops) to thrive and thwarting species which might otherwise crowd them out (weeds). In essence, with agriculture we are creating a very simple ecosystem. At most, it has only three levels-producers (crops), primary consumers (livestock, humans) and secondary consumers (humans). This means that little energy is lost between tropic levels, since there are fewer trophic levels present. This is good for humans, but what type of "ecosystem" have we created? Agricultural ecosystems have several problems. First, we create *monocultures*, or fields with only one crop. This is simplest for planting, weeding, and harvesting, but it also packs many similar plants into a small area, creating a situation ideal for disease and insect pests. In natural ecosystems, plants of one species are often scattered. Insects, which often specialize on feeding on a particular plant species, have a hard time finding the scattered plants. Without food, the insect populations are kept in check. In a field of corn however, even the most inept insect can find a new host plant with a jump in any direction. Likewise, disease is more easily spread if the plants are in close proximity. It takes lots of chemicals (pesticides) to keep a monoculture going.

Another problem with human agriculture is that we rely on relatively few plants for food. If the corn and rice crops failed worldwide in the same year, we would be hard-pressed to feed everyone. tural ecosystems usually have alternate sources of food available if one fails. A final problem associated with agroecosystems is the problem of inorganic nutrient recycling.

In a natural ecosystem, when a plant dies it fall to the ground and rots, and its inorganic nutrients are returned to the soil from which they were taken. In human agriculture, however, we harvest the crop, truck it away, and flush it down the toilet to be run off in the rivers to the ocean. Aside from the water pollution problems this causes, it should be obvious to you that the nutrients are not returned to the fields. They have to be replaced with chemical fertilizers, and that means mining, transportation, electricity, etc. Also, the chemical fertilizers tend to run off the fields (along with soil disrupted by cultivation) and further pollute the water.

Some solutions are at hand, but they bring on new problems, too. No-till farming uses herbicides to kill plants in a field; the crop is then planted through the dead plants without plowing up the soil. This reduced soil and fertilizer erosion, but the herbicides themselves may damage ecosystems. In many areas, sewage sludge is returned to fields to act as a fertilizer.

This reduces the need for chemical fertilizers, but still requires a lot of energy to haul the sludge around. Further, if one is not careful, things such as household chemicals and heavy metals may contaminate the sewage sludge and biomagnify in the crops which we would then eat.

BioGeoChemical Cycles

We have already seen that while energy does not cycle through an ecosystem, chemicals do. The inorganic nutrients cycle through more than the organisms, however, they also enter into the atmosphere, the oceans, and even rocks. Since these *chem*icals cycle through both the *bio*logical and the *geo*logical world, we call the overall cycles biogeochemical cycles. Each chemical has its own unique cycle, but all of the cycles do have some things in common. *Reservoirs* are those parts of the cycle where the chemical is held in large quantities for long periods of time. In *exchange pools*, on the other hand, the chemical is held for only a short time. The length of time a chemical is held in an exchange pool or a reservoir is termed its *residence* time. The oceans are a reservoir for water, while a cloud is an exchange pool. Water may reside in an ocean for thousands of years, but in a cloud for a few days at best. The biotic community includes all living organisms. This community may serve as an exchange pool (although for some chemicals like carbon, bound in a sequoia for a thousand years, it may seem more like a reservoir), and also serve to move chemicals from one stage of the cycle to another.

For instance, the trees of the tropical rain forest bring water up from the forest floor to be evapourated into the atmosphere. Likewise, coral endosymbionts take carbon from the water and turn it into limestone rock. The energy for most of the transportation of chemicals from one place to another is provided either by the sun or by the heat released from the mantle and core of the Earth. While all inorganic nutrients cycle, we will focus on only 4 of the most important cycles- water, carbon (and oxygen), nitrogen, and phosphorous.

The Water Cycle

In the water cycle, energy is supplied by the sun, which drives evapouration whether it be from ocean surfaces or from treetops. The sun also provides the energy which drives the weather systems which move the water vapour (clouds) from one place to another (otherwise, it would only rain over the oceans). Precipitation occurs when water condenses from a gaseous state in the atmosphere and falls to earth. Evapouration is the reverse process in which liquid water becomes gaseous.

Gravity continues to operate, either pulling the water underground (groundwater) or across the surface (run-off). In either event, gravity continues to pull water lower and lower until it reaches the oceans (in most cases; the Great Salt Lake, Dead Sea, Caspian Sea, and other such depressions may also serve as the lowest basin into which water can be drawn). Frozen water may be trapped in cooler regions of the Earth (the poles, glaciers on mountaintops, etc.) as snow or ice, and may remain as such for very long periods of time. Lakes, ponds, and wetlands form where water is temporarily trapped.

The oceans are salty because any weathering of minerals that occurs as the water runs to the ocean will add to the mineral content of the water, but water cannot leave the oceans except by evapouration, and evapouration leaves the minerals behind. Thus, rainfall and snowfall are comprised of relatively clean water, with the exception of pollutants (such as acids) picked up as the waster falls through the atmosphere. Organisms play an important role in the water cycle.

As you know, most organisms contain a significant amount of water. This water is not held for any length of time and moves out of the organism rather quickly in most cases. Animals and plants lose water through evapouration from the body surfaces, and through evapouration from the gas exchange structures (such as lungs).

In plants, water is drawn in at the roots and moves to the gas exchange organs, the leaves, where it evapourates quickly. This special case is called transpiration because it is responsible for so much of the water that enters the atmosphere.

In both plants and animals, the breakdown of carbohydrates (sugars) to produce energy (respiration) produces both carbon dioxide and water as waste products. Photosynthesis reverses this reaction, and water and carbon dioxide are combined to form carbohydrates. Now you understand the relevance of the term carbohydrate; it refers to the combination of carbon and water in the sugars we call carbohydrates.

Carbon Cycle

Once you understand the water cycle, the carbon cycle is relatively simple. From a biological perspective, the key events here are the complementary reactions of respiration and photosynthesis. Respiration takes carbohydrates and oxygen and combines them to produce carbon dioxide, water, and energy.

Photosynthesis takes carbon dioxide and water and produces carbohydrates and oxygen. Photosynthesis takes energy from the sun and stores it in the carbon-carbon bonds of carbohydrates; respiration releases that energy. Both plants and animals carry on respiration, but only plants (and other producers) can carry on photosynthesis. The chief reservoirs for carbon dioxide are in the oceans and in rock.

Carbon dioxide dissolves readily in water. Once there, it may precipitate (fall out of solution) as a solid rock known as calcium carbonate (limestone). Corals and algae encourage this reaction and build up limestone reefs in the process. On land and in the water, plants take up carbon dioxide and convert it into carbohydrates through photosynthesis.

This carbon in the plants now has 3 possible fates. It can be liberated to the atmosphere by the plant through respiration; it can be eaten by an animal, or it can be present in the plant when the plant dies. Animals obtain all their carbon in their food, and, thus, all carbon in biological systems ultimately comes from plants (autotrophs).

In the animal, the carbon also has the same 3 possible fates. Carbon from plants or animals that is released to the atmosphere through respiration will either be taken up by a plant in photosynthesis or dissolved in the oceans. When an animal or a plant dies, 2 things can happen to the carbon in it. It can either be respired by decomposers (and released to the atmosphere), or it can be buried intact and ultimately form coal, oil, or natural gas (fossil fuels).

The fossil fuels can be mined and burned in the future; releasing carbon dioxide to the atmosphere. Otherwise, the carbon in limestone or other sediments can only be released to the atmosphere when they are subducted and brought to volcanoes, or when they are pushed to the surface and slowly weathered away. Humans have a great impact on the carbon cycle because when we burn fossil fuels we release excess carbon dioxide into the atmosphere.

This means that more carbon dioxide goes into the oceans, and more is present in the atmosphere. The latter condition causes global warming, because the carbon dioxide in the atmosphere allows more energy to reach the Earth from the sun than it allows to escape from the Earth into space.

The Oxygen Cycle

If you look back at the carbon cycle, we have also described the oxygen cycle, since these atoms often are combined. Oxygen is present in the carbon dioxide, in the carbohydrates, in water, and as a molecule of two oxygen atoms. Oxygen is released to the atmosphere by autotrophs during photosynthesis and taken up by both autotrophs and heterotrophs during respiration.

In fact, all of the oxygen in the atmosphere is *biogenic*; that is, it was released from water through photosynthesis by autotrophs. It took about 2 billion years for autotrophs (mostly cyanobacteria) to raise the oxygen content of the atmosphere to the 21% that it is today; this opened the door for complex organisms such as multicellular animals, which need a lot of oxygen.

The Nitrogen Cycle

The nitrogen cycle is one of the most difficult of the cycles to learn, simply because there are so many important forms of nitrogen, and because organisms are responsible for each of the interconversions. Remember that nitrogen is critically important in forming the amino portions of the amino acids which in turn form the proteins of your body. Proteins make up skin and muscle, among other important structural portions of your body, and all enzymes are proteins.

Since enzymes carry out almost all of the chemical reactions in your body, it's easy how important nitrogen is. The chief reservoir of nitrogen is the atmosphere, which is about 78% nitrogen. It is a pretty non-reactive gas; it takes a lot of energy to get nitrogen gas to break up and combine with other things, such as carbon or oxygen. Nitrogen gas can be taken from the atmosphere (fixed) in two basic ways. First, lightning provides enough energy to "burn" the nitrogen and fix it in the form of nitrate, which is a nitrogen with three oxygens attached. This process is duplicated in fertilizer factories to produce nitrogen fertilizers.

These nitrogen-fixing bacteria come in three forms: some are free-living in the soil; some form symbiotic, mutualistic associations with the roots of bean plants and other legumes (rhizobial bacteria); and the third form of nitrogen-fixing bacteria are the photosynthetic cyanobacteria (blue-green algae) which are found most commonly in water. All of these fix nitrogen, either in the form of nitrate or in the form of ammonia (nitrogen with 3 hydrogens attached).

Most plants can take up nitrate and convert it to amino acids. Animals acquire all of their amino acids when they eat plants (or other animals). When plants or animals die (or release waste) the nitrogen is returned to the soil. The usual form of nitrogen returned to the soil in animal wastes or in the output of the decomposers, is ammonia. Ammonia is rather toxic, but, fortunately there are nitrite bacteria in the soil and in the water which take up ammonia and convert it to nitrite, which is nitrogen with two oxygens.

Nitrite is also somewhat toxic, but another type of bacteria, nitrate bacteria, take nitrite and convert it to nitrate, which can be taken up by plants to continue the cycle. We now have a cycle set up in the soil (or water), but what returns nitrogen to the air? It turns out that there are denitrifying bacteria which take the nitrate and combine the nitrogen back into nitrogen gas.

The nitrogen cycle has some important practical considerations, as anyone who has ever set up a saltwater fish tank has found out. It takes several weeks to set up such a tank, because you must have sufficient numbers of nitrite and nitrate bacteria present to detoxify the ammonia produced by the fish and decomposers in the tank. Otherwise, the ammonia levels in the tank will build up and kill the fish. This is usually not a problem in freshwater tanks for two reasons. One, the pH in a freshwater tank is at a different level than in a saltwater tank. At the pH of a freshwater tank, ammonia is not as toxic. Second, there are more multicellular plant forms that can grow in freshwater, and these plants remove the ammonia from the water very efficiently. It is hard to get enough plants growing in a saltwater tank to detoxify the water in the same way.

The Phosphorous Cycle

The phosphorous cycle is the simplest of the cycles. Phosphorous has only one form, phosphate, which is a phosphorous atom with 4 oxygen atoms. This heavy molecule never makes its way into the atmosphere, it is always part of an organism, dissolved in water, or in the form of rock. When rock with phosphate is exposed to water (especially water with a little acid in it), the rock is weathered out and goes into solution. Autotrophs take this phosphorous up and use it in a variety of ways. It is an important constituent of cell membranes, DNA, RNA, and, of course ATP, which, after all, stands for adenosine tri*phosphate*. Heterotrophs (animals) obtain their phosphorous from the plants they eat, although one type of heterotroph, the fungi, excel at taking up phosphorous and may form mutualistic symbiotic relationships with plant roots.

These relationships are called *mycorrhizae*; the plant gets phosphate from the fungus and gives the fungus sugars in return. Animals, by the way, may also use phosphorous as a component of bones, teeth and shells. When animals or plants die (or when animals defecate), the phosphate may be returned to the soil or water by the decomposers. There, it can be taken up by another plant and used again. This cycle will occur over and over until at last the phosphorous is lost at the bottom of the deepest parts of the ocean, where it becomes part of the sedimentary rocks forming there. Ultimately, this phosphorous will be released if the rock is brought to the surface and weathered. Two types of animals play a unique role in the phosphorous cycle.

The phosphate is then used as fertilizer. This mining of phosphate and use of the phosphate as fertilizer greatly accelerates the phosphorous cycle and may cause local overabundance of phosphorous, particularly in coastal regions, at the mouths of rivers, and anyplace where there is a lot of sewage released into the water (the phosphate placed on crops finds its way into our stomachs and from there to our toilets). Local abundance of phosphate can cause overgrowth of algae in the water; the algae can use up all the oxygen in the water and kill other aquatic life.

This is called eutrophication. The other animals that play a unique role in the phosphorous cycle are marine birds. These birds take phosphorous containing fish out of the ocean and return to land, where they defecate. Their guano contains high levels of phosphorous and in this way marine birds return phosphorous from the ocean to the land. The guano is often mined and may form the basis of the economy in some areas.

Energy Transformations and Biogeochemical Cycling

The study of ecosystems mainly consists of the study of certain processes that link the living, or biotic, components to the non-living, or abiotic, components. *Energy transformations* and *biogeochemical cycling* are the main processes that comprise the field of ecosystem ecology. As we learned earlier, ecology generally is defined as the interactions of organisms with one another and with the environment in which they occur. We can study ecology at the level of the individual, the population, the community, and the ecosystem. Studies of *individuals* are concerned mostly about physiology, reproduction, development or behaviour, and studies of *populations* usually focus on the habitat and resource needs of individual species, their group behaviours, population growth, and what limits their abundance or causes extinction. Studies of *communities* examine how populations of many species interact with one another, such as predators and their prey, or competitors that share common needs or resources.

In *ecosystem ecology* we put all of this together and, insofar as we can, we try to understand how the system operates as a whole. This means that, rather than worrying mainly about particular species, we try to focus on major functional aspects of the system. These *functional aspects* include such things as the amount of energy that is produced by photosynthesis, how energy or materials flow along the many steps in a food chain, or what controls the rate of decomposition of materials or the rate at which nutrients are recycled in the system.

The Transformation of Energy

The transformations of energy in an ecosystem begin first with the input of energy from the sun. Energy from the sun is captured by the process of photosynthesis. Carbon dioxide is combined with hydrogen (derived from the splitting of water molecules) to produce carbohydrates (CHO). Energy is stored in the high energy bonds of adenosine triphosphate.

The prophet Isaah said "all flesh is grass", earning him the title of first ecologist, because virtually all energy available to organisms originates in plants. Because it is the first step in the production of energy for living things, it is called *primary production Herbivores* obtain their energy by consuming plants or plant products, *carnivores* eat herbivores, and *detritivores* consume the droppings and carcasses of us all. portrays a simple food chain, in which energy from the sun, captured by plant photosynthesis, flows from *trophic level* to trophic level via the *food chain*. A trophic level is composed of organisms that make a living in the same way, that is they are all *primary producers* (plants), *primary consumers* (herbivores) or *secondary consumers* (carnivores). Dead tissue and waste products are produced at all levels.

Scavengers, detritivores, and decomposers collectively account for the use of all such "waste"—consumers of carcasses and fallen leaves may be other animals, such as crows and beetles, but ultimately it is the microbes that finish the job of decomposition. Not surprisingly, the amount of primary production varies a great deal from place to place, due to differences in the amount of solar radiation and the availability of nutrients and water. Energy transfer through the food chain is inefficient.

This means that less energy is available at the herbivore level than at the primary producer level, less yet at the carnivore level, and so on. The result is a pyramid of energy, with important implications for understanding the quantity of life that can be supported. Usually when we think of food chains we visualize green plants, herbivores, and so on. These are referred to as *grazer food chains*, because living plants are directly consumed. In many circumstances the principal energy input is not green plants but dead organic matter.

These are called *detritus food chains*. Examples include the forest floor or a woodland stream in a forested area, a salt marsh, and most obviously, the ocean floor in very deep areas where all sunlight is extinguished 1000's of meters above. In subsequent lectures we shall return to these important issues concerning energy flow. Finally, although we have been talking about food chains, in reality the organization of biological systems is much more complicated than can be represented by a simple "chain". There are many food links and chains in an ecosystem, and we refer to all of these linkages as a *food web*. Food webs can be very complicated, where it appears that *"everything is connected to everything else"*, and it is important to understand what are the most important linkages in any particlar food web.

Biogeochemistry

One obvious way is to study the flow of energy or the cycling of elements. The cycling of elements is controlled in part by organisms, which store or transform elements, and in part by the chemistry and geology of the naturalworld. The term *Biogeochemistry* is defined as the study of how living systems influence, and are controlled by, the geology and chemistry ofthe earth. Thus biogeochemistry encompasses many aspects of the abiotic and biotic world that we live in.

There are several main *principles and tools* that biogeochemists use to study earth systems. Most of the major environmental problems that we face in our world toady can be analysed using biogeochemical principles and tools. These problems include global warming, acid rain, environmental pollution, and increasing greenhouse gases. The principles and tools that we use can be broken down into 3 major components: *element ratios, mass balance, and element cycling.*

Element Ratios

In biological systems, we refer to important elements as *"conservative"*. These elements are often nutrients. By "conservative" we mean that an organism can change only slightly the amount of these elements in their tissues if they are to remain in good health. It is easiest to think of these conservative elements in relation to other important elements in the organism. In healthy algae the elements C, N, P, and Fe have the following ratio, called the *Redfield ratio* after the oceanographer who discovered it:

C: N: P: Fe = 106: 16: 1: 0.01

Once we know these ratios, we can compare them to the ratios that we measure in a sample of algae to determine if the algae are lacking in one of these limiting nutrients.

Mass Balance

Another important tool that biogeochemists use is a simple mass balance equation to describe the state of a system. The system could be a snake, a tree, a lake, or the entire globe. Using a mass balance approach we can determine whether the system is changing and how fast it is changing.

The equation is:

Net Change = Input + Output + Internal Change

In this equation the net change in the system from one time period to another is determined by what the inputs are, what the outputs are, and what the internal change in the system was. The example given in class is of the acidification of a lake, considering the inputs and outputs and internal change of acid in the lake.

Element Cycling

Element cycling describes where and how fast elements move in a system. There are two general classes of systems that we can analyse. A closed system refers to a system where the inputs and outputs are negligible compared to the internal changes. Examples of such systems would include a bottle, or our entire globe. There are two ways we can describe the cycling of materials within this closed system, either by looking at the rate of movement or at the pathways of movement.

- Rate = number of cycles/time * as rate increases, productivity increases
- Pathways-important because of different reactions that may occur

In an open system there are inputs and outputs as well as the internal cycling. Thus we can describe the rates of movement and the pathways, just as we did for the closed system, but we can also define a new concept called the residence time. The residence time indicates how long on average an element remains within the system before leaving the system.

- Rate
- Pathways
- Residence time, Rt
- Rt = Total amount of matter/output rate of matter

Controls on Ecosystem Function

Now that we have learned something about how ecosystems are put together and how materials and energy flow through ecosystems, we can better address the question of "what controls ecosystem function"? There are two dominant theories of the control of ecosystems. The first, called bottom-up control, states that it is the nutrient supply to the primary producers that ultimately controls how ecosystems function.

If the nutrient supply is increased, the resulting increase in production of autotrophs is propagated through the food web and all of the other trophic levels will respond to the increased availability of food (energy and materials will cycle faster). The second theory, called top-down control, states that predation and grazing by higher trophic levels on lower trophic levels ultimately controls ecosystem function.

If you have an increase in predators, that increase will result in fewer grazers, and that decrease in grazers will result in turn in more primary producers because fewer of them are being eaten by the grazers. Thus the control of population numbers and overall productivity "cascades" from the top levels of the food chain down to the bottom trophic levels. So, which theory is correct? Well, as is often the case when there is a clear dichotomy to choose from, the answer lies somewhere in the middle.

There is evidence from many ecosystem studies that both controls are operating to some degree, but that neither control is complete. The "top-down" effect is often very strong at trophic levels near to the top predators, but the control weakens as you move further down the food chain. Similarly, the "bottom-up" effect of adding nutrients usually stimulates primary production, but the stimulaton of secondary production further up the food chain is less strong or is absent.

Thus we find that both of these controls are operating in any system at any time, and we must understand the relative importance of each control in order to help us to predict how an ecosystem will behave or change under different circumstancs, such as in the face of a changing climate.

The Geography of Ecosystems

There are many different ecosystems: rain forests and tundra, coral reefs and ponds, grasslands and deserts. Climate differences from place to place largely determine the types of ecosystems. How terrestrial ecosystems appear to us is influenced mainly by the dominant vegetation. The word "biome" s used to describe a major vegetation type such as tropical rain forest, grassland, tundra, etc., extending over a large geographic area. It is neve used for aquatic systems, such as ponds or coral reefs. It always refers to a vegetation category that is dominant overa very large geographic scale, and so is somewhat broader than an ecosystem. Every place on earth gets the same total number of hours of sunlight each year, but not the same amount of heat.

The sun's rays strike low latitudes directly but high latitudes obliquely. This uneven distribution of heat sets up not just temperature differences, but global wind and ocean currents that in turn have a great deal to do with where rainfall occurs. Add in the cooling effects of elvation and the effects of land masses on temperature and rainfall, and we get a complicated global pattern of climate. A schematic view of the earth shows that, complicated though climate may be, many apects are predictable High solar energy striking near the equator ensures nearly constant high temperatures and high rates of evapouration and plant transpiration.

Warm air rises, cools, and sheds its moisture, creating just the conditions for a tropical rain forest. Contrast the stable temperature but varying rainfall of a site in Panama with the relatively constant precipitation but seasonally hanging temperature of a site in New York State. Every location has a rainfall- temperature graph that is typical of a broader region.

We can draw upon plant physiology to know that certain plants are distinctive of certain climates, creating te vegetation appearance that we call biomes. Note how well the distribution of biomes plots on the distribution of climates Note also that some climates are ipossible, at least on our planet. High precipitation is not possible at low temperatures—there is not enough solar energy to power the water cycle, and most water is frozen and thus biologically unavailable throughout the year.

The high tundra is as much a desert as is the Sahara:

- Ecosystems are made up of abiotic (non-living, environmental) and biotic components, and these basic components are important to nearly all tpes of ecosystems. Ecosystem Ecology looks at energy transformations and biogeochemical cycling within ecosystems.
- Energy is continually input into an ecosystem in the form of light energy, and some energy is lost with each transfer to a higher trophic level. Nutrients, on the other hand, are recycled within an ecosystem, and their supply normally limits biological activity. So, "energy flows, elements cycle".
- Energy is moved through an ecosystem via a food web, which is made up of interlocking food chains. Energy is first captured

by photosynthesis (primary poduction). The amount of primary production determines the amount o energy available to higher trophic levels.

- The study of how chemical elements cycle through an ecosyste is tered biogeochemistry. A biogeochemical cycle can be expressed as a set of stores (pools) and transfers, and can be studied using the concepts of "stoichiometry", "mass balance", and "residence time".
- Ecosystem function is controlled mainly by two processes, "top-down" and "bottom-up" controls.
- A biome is a major vegetation type extending over a large area. Biome distributions are determined largely by temperature and precipitation patterns on the Earth's surfae.

Biogeochemistry

As the terminology implis, Biogeochemistry is a borderline specialisation involving three basic branches of natural, physical and earth sciences. On one hand living organisms are involved and on the other substances such asnutrients, are made available to the living organisms either through soil (the pedosphere) or through water (the hydrosphere). Of course there arecertain organisms which can get their food through air (atmosphere) also.

The living organisms, while differing vastly among themselves in size, morphology and physiology, share, all the same a basic feature common to them all: ability to metabolise their requirements in the habitats they live. An individual organism is perishable but life process, sustained by successive generation, is perpetual Biogeochemistry is all about impact of such living processes on our environment. The bulk of living matter is concentrated in the landmass of the world though at the continent-ocean interface and at the bottom of the sea they are present in abundance. The physio-chemical interaction between organisms and the land that sustain them have been a subject of study for over a century now.

More than two hundred years ago, The turnover of elements on the surface of the globe in hich he postulated on the cyclic exchange of elements among the three kingdoms of nature, namely mineral, vegetable and animal. Lavoisier paid a heavy price for stating so: he was executed after his work was published. The interaction among the above three kingdoms essentially constitute the broad field o Biogeochemistry as we know today.

The original kingdoms have been replaced by reservoirs of different kind and atmosphere is brought into the picture because of gases involved in many biogeochemical processes. Since life is cyclic, all processes where living matter is involved has also got to be cyclic. Hence, in all the reservoirs are linked to each other in both directions:at the inlet and outlet point for each reservoir.

The interconnection of all the reservoirs leads to a overall global cycle. Such a cycle is known by various names depending on the user and the sector of interest: biogeochemical cycle, hydrological cycle, geochemical cycle and rock cycle.

The emphasis in each cycle varies: in hydrological cycle, one will be interested in transfer of water; in the geochemical cycle, one will be interested in transfer from the crust to other segments and in rock cycle, the interest will be transformation of large volume of crustal materials across various interfaces. Whereas in the biogeochemical cycle the focus is on nutrients essential for the life processes and at the toxic substances determinant to such living things.

Thus biogeochemical processes can be defined in three stages:

- *Stage*-1: Elements to be grouped according to the needs: Essential, such as Carbon (C), Nitrogen (N) and Phosphorous (P) Minor, such as Sodium (Na), Potassium(K) etc.

  Trace, such as Iron (Fe), Manganese (Mn), Copper (Cu) etc.

  Toxic, such as Mercury (Hg), Selenium (Se), Cadmium (Cd) (all the above groupings according to the requirements of biological system in the biological world)-the bio part of the topic.
- *Stage*-2: The availability of the above from the earth, generally from the top few metres of the earth (the geocomponent, more specifically the pedosphere the geo part of the topic and now.
- *Stage*-3: The physio-chemical mechanisms that make the needs of the biological systems from the pedospheric layers of the earth in response to micro-environmental functions such as pH (acidity), weathering and release of elements by their breakdown due to chemical interaction of individual minerals with water. This constitutes the chemical part of the topic.

Let this linkages using an essential element Carbon as an example. Referring to carbon is present in the atmosphere mainly as carbon dioxide (CO_2) and in small quantities as carbon monoxides (CO). In the hydrosphere, carbon is present as bicarbonate and carbonate the major factor in the hardness of water. In the lithosphere- pedosphere, it is present as calcium carbonate ($CaCO_3$) or some combination thereof. In the biosphere, carbon is present as the building block of the plants and animals in the form of variety of carbon compounds suh as carbohydrates, proteins and amino-acids. Carbon enters the biological system either through geological process (weathering of carbon-bearing minerals such as limestone ($CaCO_3$) and/or chemical process directly from the atmosphere (CO_2) by photosynthesis.

When the organism (including plant) dies out, the carbon is released back either directly to the atmosphere as CO_2 or is mineralised as coal, limestone etc. So that it is returned to the lithosphere. Thus the carbon cycle is complete.

The oldest carbon-bearing organic world remains known to man is about 3.5 billion years whereas the age of the earth is at present estimated to be about 4.5 billion years. In spite of the complexity of the biological evolution through geological time scale, carbon has been and is being cycled through various reservoirs relatively unaffected and un- disturbed till recent times.

Among the three essential nutrients for life phosphorous (P) is extremely interesting. Unlike carbon and nitrogen, it is predominately derived from the crustal top of the solid earth and mobilised through the biosphere by the transport behaviour in the hydrosphereP ignites in air but glows when left in darkness. Thus it is a carrier of energy in all living systems.

The homologous elements P and N has complimentary properties in relation to Hydrogen(H) and Oxygen(O_2)- the other important elements for most living matter on the surface of the earth. P in association with O_2 introduces some structural order in the molecules at cellular levels and thus the P-O association in the form of PO_4 units separates life from ordinary inanimate minerals represented by similar association of Silicon with O_2 (SiO_4)

From a single cell amoeba some 3.5 billion years ago (3500 million), organisms progressively went through complex biological evolution as a function of geological time scale.

The lin between time and evolution is established through changing environmental parameters such the carbon dioxide (CO_2) levels in atmosphere prevailing at a given time, the O_2 levels, nutrients such as P and minor and trace elements etc. Geochemical evidences through rock records show that about 2.5 billion years ago, the atmospheric CO_2 was at least 100 times lower as that of today and O_2 levels negligible compared to today s environment.

Have been continuously interacting among each other through such a vast time scale to maintain the cyclic behaviour of nutrients and other elements. Excessive constituents in any one segment was removed over a period of time to other segments so that its biogeochemical cyclic behaviour was not affected.

During Gondwana times (about 400 to 500 million years before) when India, Australia and Antarctica were neighbours, excess CO_2 in the atmosphere was neutralised by enhanced photosynthesis and subsequently the biota decayed and gave raise to most of the coal deposits which we are exploiting today. Thus due to the biogeochemical behaviour, nature was able to adjust itself at its own rate in any parameters.

On the other hand in modern times, man introduces any one variable to a segment in very abruptly and in such large dose that nature is unable to adjust to that change. A case in point the emission of green house gas CO_2 by the burning of coal. The essential nutrient element C is transferred from the lithosphere to the atmosphere so fast that the excess CO_2 is unable to be removed equally faster by the hydrosphere leading to enhanced levels of CO_2 in the atmosphere and hence we hear everyday about the global warming On the other hand, let us take a element such as Mercury (Hg) considered to be very toxic to the biological world. The movement of Hg in cyclic fashion through our environment. Hg generally occurs in minerals in association with sulphur in the form of mercury sulphide (HgS) systems.

Even in rocks, it is considered to be trace element since its abundance in the crustal earth is very low. However, during the exploitation of base metals resources, such as that of copper, lead and Zinc, mercury is extracted as a buy-product. Modern civilization uses mercury for a variety of products such as chemicals, coatings on seeds etc. Even though it does not react with water in the hydrosphere easily, mercury combines with any organic substances such decayed plant debris etc and immediately becomes toxic to the living biota.

The well known minemata decease in Japan was the result of such affinity of mercury for many biotic materials. Being the only metal known to man to be in liquid state at room conditions, it easily vapourises and hence get also into the atmospheric segment Hence due to man s use, what was once locked up safely in inert geological materials in rocks, have been mobilised to the biosphere, hydrosphere and atmosphere. The story is basically same for a few other toxic metals such as Cadmium though the details may individually vary due to their differing biogeochemical behaviour in our present environment. It is clear that processes in our environment are better understood by focussing on the interfacing areas.

Thus Biogeochemistry is the ideal such frontier area of study and research. It is however a relatively new field and only in the last two decades various researchers across the world have been focussing their attention on this inter- disciplinary subject.

Outside India there are a few places where separate institutes exists for this purpose. At about the same time the Ministry of Environment and Forests also approved the setting up of a ENVIS centre in JNU on the same subject under his direction. Though there no other specific institutions in India in this subject, there are individual researchers in many universities working on topics of interest to Biogeochemistry.

Thus, the area of Biogeochemistry is an emerging branch of multi-diciplinary approach involving all major branches of Sciences. From a humble beginning two decades ago, now this subject is well established in the international field and in India is fast catching up as an important tool of research for understanding processes in our environment. Sufficient infra- structure has now developed in India through a variety of individual and institutional efforts so that in the years to come the importance work on Biogeochemistry will be better understood and appreciated.

The involvement of a number of specialists is obvious when ones looks into the above three stages: for understanding the biotic behaviour, one should have a good understanding of ecological principles; the requirements for the ecosystem is provided by basic geochemical processes of weathering helped by microbiological and biochemical reactions in soils and over rocks; farther, from the source to the receptor, chemical and hydrological principles operate as a middleman.

Hence within the broad field of Biogeochemistry itself, there are sub-specialisations depending on the background of the researchers- be a ecologist, geochemist or chemical hydrologist. Hence the word Biogeochemistry is truly an integrating field to better understand our present, past and future environmental problems.

Among the more recent sub-specialisations within Biogeochemistry that have emerged in the international research is the application of certain biochemical and molecular biochemical techniques to understand our present and past environment. A full human body was found in tact in the frozen alps and several investigations on the body using most modern biochemical techniques have thrown some light about the habitat of the community some 5000 years ago, then the prevailing eco- system, palaeo-environment, anthro-pological lineage etc. DNA based study thus offers excellent scope even in environment related matter.

In India, a student from our labouratory is working under the supervision of Prof. Laljit Singh, Centre for DNA Finger-printing at Hyderabad to apply DNA techniques in fossils bones of geological and anthropological importance to understand the habitat and palaeo-environment. DNA techniques to ancient biological specimens. Farther, characterisations of amino-acids in ancient sediments have been successfully used to understand palaeo-environment and also stresses acting on such systems.

Thus, Biogeochemistry can be viewed as a very challenging area of study involving a variety of basic sciences so that it is ideally suited to understand our environment better. Research should not go in the direction of reinventing the wheel but improve and if possible, replace it. Hopefully the 21st century consolidation of the field of Biogeochemistry as an established brach of science.

Enlarged version of specific component. whenevr possible. Important components in each segment and direction of transfer F means flux that is concentration for a given parameter the volume of water/air transferred in that direction per unit time. This, F has unit of wt/vol/time. F is likely to have volumes since the time involved is relatively shorter for some of the sectors. For eg. some of the variables transferred involving soil and plants/animals and plants and atmosphere etc. may have a time scale of a few seconds only.

The carbon transfer, the following fluxes have been incorporated: 1 to 34 point all have same magnitude, time scale and importance. Between river and young soils sediments, the transfer 26 and 27 represent deposition and erosion processes. Faster compared to deposition or accumulation so that net transfer is loss of carbon from soil (Pedosphere) to river (hydrosphere).

Hg cycle is shwn in tow stages: pre-man and present day. this concept is adapted from Garrels *et al.,* 1976. The main differences is indicated in teh Hg-mining in the present day cause additional flux-vaporisation-from land to atmosphere. ++ or + singns indicate higher gains in the respectinve sectors the present day cycle relative to pre-man cycle. As a consequence thee is also poportional increase in the flxes in the present day cycle between the concerned sectors. Garels et al. (1976) estimate that Hg will be the first metal that is likely to be lost out entiely ot of the solid spheres (litho-and pedospheres) to the fluid (hydro- and atmo-) spheres in foreseabe future.

Mercury in the Environment

Mercury is a silvery, liquid metal at room temperature and is often referred to as one of the "heavy metals." Like water, mercury can evapourate and become airborne. Because it is an element, mercury does not break down into less toxic substances. Once mercury escapes to the environment, it circulates in and out of the atmosphere until it ends up in the bottoms of lakes and oceans. Mercury can be found as the elemental metal or in a wide variety of organic and inorganic compounds. Depending on its chemical form, mercury may travel long distances before it falls to earth with precipitation or dust. Bacteria and chemical reactions in lakes and wetlands can change the mercury into a much more toxic form known as methylmercury. Fish become contaminated with methylmercury by eating food (plankton and smaller fish), which has absorbed methylmercury.

As long as the fish continue to be exposed to mercury, mercury continually builds up in fish's bodies. Fish that eat other fish become even more highly contaminated. Thus, the largest tend to be the most contaminated.

When people eat the contaminated fish, the methylmercury can remain in their bodies for a long time. If they eat fish containing methylmercury faster than their bodies can get discharge it, the methylmercury accumulates in their bodies and can be toxic. Many states have fish consumption advisories to inform people about how many meals of fish they can safely eat over a period of time.

Mercury's Environmental Effects

Fish are the main source of food for many birds and other animals, and mercury can seriously damage the health of these species. Loons, eagles, panthers, otters, mink, kingfishers and ospreys naturally eat large quantities of fish. Because these predators rely on speed and coordination to obtain food, mercury may be particularly hazardous to these animals.

Recent research in Minnesota indicates that the following environmental effects are occurring:

- Loons are accumulating so much mercury that it may be affecting their ability to reproduce;

- Elevated levels of mercury have been found in mink and otters;
- Walleye reproduction may be impaired by the fish's exposure to mercury.

Similar effects are being documented for other fish and fish-eating species around the United States and Canada. Has there always been mercury contamination, or is this a recent problem? This is a difficult question to answer, in part because of a lack of adequately preserved fish specimens of preindustrial age to compare against contemporary samples. However, several lines of evidence from recent studies on Wisconsin lakes suggest that increased emissions to the atmosphere, and subsequent higher deposition rates to lakes, likely translate into higher mercury levels in fish.

The Mercury Cycle and Bioaccumulation

There is a constant biogeochemical cycle of mercury. This cycle includes:

- Release of elemental mercury as a gas from the rocks and waters (degassing);
- Long-range transport of the gases in the atmosphere;
- Wet and dry deposition upon land and surface water;
- Absorption onto sediment particles;
- Bioaccumulation (or biomagnification) in terrestrial and aquatic food chains.

Bioaccumulation means an increase in the concentration of a chemical in an organism over time, compared to the chemical's concentration in the environment. Bioaccumulation can be a normal and essential process for the growth of any species, but the accumulation of unnecessary chemicals or toxins, or even the overaccumulation of essential substances can be detrimental.

All animals, including humans, daily bioaccumulate many vital nutrients, such as vitamins A, D, and K, trace minerals, essential fats and amino acids, but unfortunately, they can also accumulate many unnecessary substances, such as lead or mercury. What concerns toxicologists is the bioaccumulation of necessary substances to levels in the body that can cause harm.

With substances such as lead or mercury, any accumulation at all can be harmful. Compounds accumulate in living things any time they are taken up and stored faster than they are broken down (metabolized) or excreted. Understanding the dynamic process of bioaccumulation is important in protecting humans and other organisms from the adverse effects of chemical exposure, and it has become a critical consideration in the regulation of chemicals.

Bioaccumulation varies among individual organisms as well as among species. Large, fat, long-lived individuals or species with low rates of metabolism or excretion of a chemical will bioaccumulate more than small, thin, short-lived organisms. Thus, an old lake trout may bioaccumulate much more than a young bluegill in the same lake.

Above is a schematic drawing of mercury cycling in an aquatic ecosystem. With the exception of isolated cases of known point sources, the source of most mercury to most aquatic ecosystems is deposition from the atmosphere, primarily associated with rainfall.

In the aquatic environment, mercury can be:

- Dissolved or suspended in the water
- Trapped in the sediments
- Ingested by living things (biota)

Methylmercury is the form of mercury most available and most toxic to biota (including zooplankton, insects, fish, and humans). This form of mercury is easily taken up by biota and bioaccumulates in their tissues. Unlike many other fish contaminants, such as PCBs, dioxin, and DDT, mercury does not concentrate in the fat, but in the muscle tissue. Thus, there is no simple way to remove mercury-contaminated portions from fish that is to be eaten.

Life and Biogeochemical Cycles

The Sulfur Cycle

Another example of a major biogeochemical cycle of significance to climate and life is the sulfur cycle. Living things require certain safe, low levels of this nutrient. The sulfur cycle can be thought of as beginning with the gas sulfur dioxide (SO_2) or the particles of sulfate (SO_4=) compounds in the air. These compounds either fall out or are rained out of the atmosphere. Plants take up some forms of these compounds and incorporate them into their tissues. Then, as with nitrogen, these organic sulfur compounds are returned to the land or water after the plants die or are consumed by animals. Bacteria are important here as well since they can transform the organic sulfur to hydrogen sulfide gas (H_2S). In the oceans, certain phytoplankton can produce a chemical that transforms to SO_2 that resides in the atmosphere. These gases can re-enter the atmosphere, water, and soil, and continue the cycle. In its reduced oxidation state, the nutrient sulfur plays an important part in the structure and function of proteins. In its fully oxidized state, sulfur exists as sulfate and is the major cause of acidity in both natural and polluted rainwater. This link to acidity makes sulfur important to geochemical, atmospheric, and biological processes such as the natural weathering of rocks, acid precipitation, and rates of denitrification. Sulfur is also one of the main elemental cycles most heavily perturbed by human activity. Estimates suggest that emissions of sulfur to the atmosphere from human activity are at least equal or probably larger in magnitude than those from natural processes. Like nitrogen, sulfur can exist in many forms: as gases or sulfuric acid particles. Sulfuric acid particles contribute to the polluting smog that engulfs some industrial centres and cities where many sulfur containing fuels are burned. Such particles floating in air (known as sulfate aerosols) can cause respiratory diseases or cool the climate by reflecting some extra sunlight to space.

The lifetime of most sulfur compounds in the air is relatively short (*e.g.* days). Superimposed on these fast cycles of sulfur are the extremely slow sedimentary-cycle processes or erosion, sedimentation, and uplift of rocks containing sulfur.

In addition, sulfur compounds from volcanoes are intermittently injected into the atmosphere, and a continual stream of these compounds is produced from industrial activities. These compounds mix with water vapour and form sulfuric acid smog. In addition to contributing to acid rain, the sulfuric acid droplets of smog form a haze layer that reflects solar radiation and can cause a cooling of the earth's surface. While many questions remain concerning specifics, the sulfur cycle in general, and acid rain and smog issues in particular are becoming major physical, biological, and social problems.

Biogeochemical Cycles

The Earth is a closed system for matter, except for small amounts of cosmic debris that enter the Earth's atmosphere. This means that all the elements needed for the structure and chemical processes of life come from the elements that were present in the Earth's crust when it was formed billions of years ago. This matter, the building blocks of life, continually cycle through Earth's systems, the atmosphere, hydrosphere, biosphere, and lithosphere, on time scales that range from a few days to millions of years. These cycles are called biogeochemical cycles, because they include a variety of biological, geological, and chemical processes. Many elements cycle through ecosystems, organisms, air, water, and soil. Many of these are trace elements. Other elements, including carbon, nitrogen, oxygen, hydrogen, sulfur, and phosphorus are critical components of all biological life. Together, oxygen and carbon account for 80 per cent of the weight of human beings. Because these elements are key components of life, they must be available for biological processes.

Carbon, however, is relatively rare in the Earth's crust, and nitrogen, though abundant in the atmosphere, is in a form that is not useable by living organisms.The biogeochemical cycles transport and store these important elements so that they can be used by living organisms. Each cycle takes many different pathways and has various reservoirs, or storage places, where elements may reside for short or long periods of time.

Each of the chemical, biological, and geological processes varies in their rates of cycling. Some molecules may cycle very quickly depending on the pathway. Carbon atoms in deep ocean sediments may take hundreds to millions of years to cycle completely through the system. An average water molecule resides in the atmosphere for about ten days, although it may be transported many miles before it falls back to the Earth as rain.

How fast substances cycle depends on its chemical reactivity and whether or not it can be found in a gaseous state. A gaseous phase allows molecules to be transported quickly. Phosphorous has no gaseous phase and is relatively uncreative, so it moves very slowly through its cycle. Phosphorus is stored in large amounts in sediment in the oceans or in the Earth's crust and is recycled back to the surface only over very long periods of time through upwelling of ocean waters or weathering of rocks.

Biogeochemical cycles are subject to disturbance by human activities. Humans accelerate natural biogeochemical cycles when elements are extracted from their reservoirs, or sources, and deposited back into the environment (sinks). Humans have significantly altered the carbon cycle by extracting and combusting billions of tons of hydrocarbons in fossil that were buried deep in the Earth's crust, in addition to clearing vegetation that stores carbon.

Global release of carbon through human activities has increased from 1 billion tons per year in 1940 to 6.5 billion tons per year in 2000. About half of this extra carbon is taken up by plants and the oceans, while the other half remains in the atmosphere. In addition to carbon cycle, humans have altered the nitrogen and phosphorus cycles by adding these elements to croplands as fertilizers, which has contributed to over-fertilization of aquatic ecosystems when excess amounts are carried by run-off into local waterways. Researchers are trying to understand all of the various pathways and flows of each of the biogeochemical cycles in order to understand how human activities affect these cycles.

While many important processes have been understood for more than century, there are many phenomena that scientists are just beginning to investigate. Satellite technology, among other tools, has revealed new information about interactions between the oceans and atmosphere that contribute to knowledge about the carbon cycle, but there remain many unanswered questions.

Human Modifications of Climate Services

Human activities are significantly perturbing all of these biogeochemical cycles as well as other earth system processes, both directly through industrial processes and indirectly through changing distributions and abundance of life. The atmosphere is of particular importance to the perturbations due to its crucial role in mediating all energy that enters and leaves earth. Overall, the atmosphere is the component that controls the dominant energy flow in the earth's climate system, and solar radiation from the sun provides the energy to make the weather machine work. Embedded in this process are the biogeochemical cycles we have described that operate on a variety of time and space scales and help to regulate flows of energy and materials throughout the earth system Yet, while we understand much about the functioning of separate parts of this system, there is still a great deal to be discovered about the feedbacks and linkages that allow these interconnected parts to function as a whole and, in turn, how they will respond to human modification.

Human Disturbance

Life influences the amount of CO_2 in the atmosphere through photosynthesis, respiration, and oceanic absorption. As ecosystems are altered, the balance of these processes will be altered. Human activities are upsetting this balance and increasing CO_2 in the atmosphere through the burning of fossil fuels and clearing of forests.A significant increase in CO_2 could have dramatic consequences. Mthematical models of the climate suggest that when CO_2 doubles (sometime in the middle of the next century should population, economic and technology trends continue as typically projected), the world will warm up somewhere between 1 and 5 degrees Celsius by 2100 A.D. unless other factors counteract or amplify the CO_2-induced change Even the lower end of that range is a projected warming at the rate of one degree per hundred years, a factor of ten faster than the one degree per thousand that has been the typical average rate of natural sustained global temperature change from the end of the ice ages to warmer interglacial times.

Should the higher end of the 1 to 5 degree Celsius range occur, then we could see rates of climate change some fifty times faster than sustained, natural average conditions. Climate largely determines the types of ecosystems that occupy an area. Global climate change at such a rapid rate would force many species to shift their ranges in an attempt to keep up with changing climatic conditions, as occurred during the ice age-interglacial transition ten to fifteen thousand years ago. Migrations of species such as slow growing trees with large seeds would have to occur much faster than they did in the past to keep up with rapidly shifting climates.

Other species could move more easily, raising the likelihood that communities of species could be disassembled. Estimating the rates of global warming in the next century, however, are very controversial because of the uncertainties involved with multiple interacting feedback mechanisms Humanity can control climate in ways other than changing greenhouse gas concentrations. Consider the amount of moisture released to the atmosphere through transpiration in the tropical rainforests. The dense vegetation in areas such as the Amazon basin typically recycles the precipitation that falls on it many times over, helping to form heavy cloud cover in the region.

The clouds, in turn, reflect sunlight and produce more rain, directly influencing regional climate as well as indirectly perturbing global climate through altering large-scale circulation patterns over the tropics. As humanity deforests regions like the Amazon, not only is CO_2 released into the atmosphere, but changes in the hydrologic cycle will almost certainly affect regional climate and possibly even global climatic patterns. In deforested areas of northeastern Brazil, the cutting of the tropical forests has led to desertification, changing both surface reflectivity and the rate of transpiration. This change in ecosystem character can lead to a destabilizing positive feedback, which may cause an even further reduction in precipitation.

In a recent study on the possible climatic impacts of tropical deforestation, researchers suggest that conversion of forest into crop land or pastures would cause significant changes in the local microclimate Expected changes include reduction in soil moisture, larger diurnal fluctuation of surface temperature and humidity deficit, and increased surface run-off during the rainy season and decreased run-off during the dry season. Results from general circulation model simulations of large-scale deforestation and conversion to grassy vegetation in the Amazon basin indicate an increase in surface temperature, decrease in evapotranspiration, and significant reduction in precipitation.

Depending on the scale of the disturbed areas, local climate changes can lead to regional climate changes which, in turn, may cause alterations in the global climate through atmospheric connections between tropical circulation and large-scale circulation patterns outside of the tropics. The effect on the ecological systems through changes in the hydrological cycle, an increase in the dry season, and the disruption of plant-animal interactions may make it difficult for the rainforests to re-establish themselves if they are destroyed. Climate change aside, the implications of this scenario for the conservation of biodiversity are serious.

The provision of fresh water and regulation of its flows through precipitation, evapouration, transpiration, and run off is mediated by all ecosystems. Forests and other vegetation types are critical components of this ecosystem service providing free flood and drought relief among other things. The loss of these services, through landuse change, can exacerbate disasters like spring floods in the Midwest and Southeast resulting from large expanses of land cleared for agriculture as well as the drainage of wetlands and swamps which otherwise might have acted as reservoirs for holding excess water or filtering toxic wastes.

Valuing Carbon

There is already a historic background on the evaluation of carbon that includes climate change policies such as the introduction of carbon taxes that reduce greenhouse gas emissions through increasing prices of carbon-based fuels proportional to the amount of carbon they emit Another mechanism for reducing greenhouse gas emissions is through an international tradable emissions permit system intended to limit emissions of certain pollutants. These policies and others constitute ways of balancing the economic costs of emissions with some assumed benefit of averting the loss of ecosystem services (called "climate damage").

Any comprehensive attempt to evaluate the societal value of climate change should include such things as loss of species diversity, loss of coastline from increasing sea level, environmentally displaced persons, and agricultural losses. first estimated the climate damage at 1% reduction in GNP based on market sector losses for a central estimate of climate change. This was criticized as too narrow a view of climate as a type of public good since it reflected neither non-market values (*e.g.* species loss) nor climate "surprise" scenarios conducted a survey of conventional economists, environmental economists, atmospheric scientists, and ecologists.

Their estimates of loss of gross world product (GWP) resulting from a 3 degree Celsius warming by 2090 varied between a loss of 0 and 21% of GNP with a mean of 1.9%). For a 6 degree Celsius warming scenario, the respondents predicted a loss of the world economy ranging from 0.8 to 62% with a mean estimate of 5.5%. A striking difference was noted between respondents from different academic disciplines, with natural scientists' estimates of economic impact 20 to 30 times higher than conventional economists'. Even a twoper cent loss of GWP in 1995, however, represents climate damage of hundreds of billions of dollars annually.

While it is impossible to estimate credibly a numerical value on all of the ecosystem services provided through the maintenance of the carbon cycle at present state, it may be useful to look at land use change and loss of biomass, mostly through deforestation, as a source of atmospheric CO_2. In a very simplistic and preliminary evaluation, we can use the rates of net deforestation to calculate a value for carbon. Global loss of above-ground biomass from deforestation in the tropics is approximately 1-3 gigatons/year over the past 10 years.

This amounts to between 2-5 gigatons of carbon in carbon dioxide released into the atmosphere each year from deforestation and forest degradation (this does not include the 6 gigaton carbon emissions from the burning of fossil fuels). Much of the carbon from biosphere emissions is taken up immediately by vegetation, however, leaving approximately 1-2.5 gigatons of net carbon added to the atmosphere each year. We can apply the concept of carbon taxation for emissions to an ecosystem service valuation of retaining the carbon in the forests.

Using a range of carbon taxes from typical macroeconomic models between $1 per ton and $100 per ton of carbon, the net value of carbon lost each year amounts to between $1 and $250 billion/year. However, use of optimizing economic models to estimate climate damage is highly unsatisfying since these studies use very limited and often ad hoc assumptions that both over and underestimate the likely damages to various market and non-market sectors.

Methods of Valuation

The need for alternative methods of evaluation of these climate-related ecosystem services is quite clear when examining preliminary public opinion responses of global warming. In acontroversial method called contingent valuation respondents are surveyed to determine how much they would be willing to pay to prevent a given global climate change scenario from happening or accept if so much change were to be allowed. The difficulties with this type of valuing of environmental goods and processes are immense, especially since much of the evaluation is subjective. Public opinion depends, in part, upon people's exposure to the issues and the level of education and information on these issues they have received.

In a Southern California study, the contingent valuation technique was applied to the determine the influence of potential changes in temperature and precipitation resulting from global warming on respondents willingness to pay. Factorial survey methods were used to present a variety of hypothetical climate scenarios to a sample of 600 Southern California residents. Respondents were provided with a baseline microclimate for the region before future climate scenarios were evaluated. For residents living in coastal communities, the baseline climate over the past 10 years was described as having a summer average high temperature of 75 degrees Fahrenheit, with daily highs ranging between 70 and 80 degrees, and an average of 13 inches per year of rain. One possible future scenario over the next 10 years included a summer average high temperature of 100 degrees Fahrenheit, with daily highs ranging from 80 to 120 degrees (the latter typical of Death Valley, California), and an average of 20 inches per year of rain.

With these and other scenarios, predicted probabilities were determined from the respondents willingness to pay for the abatement of different mean high temperatures. In this scenario, respondents were willing to pay an average $140 to offset a mean high temperature of 100 degrees, while a mean high temperature of 80 degrees was worth approximately $100 This represents a 40% increment in willingness to pay for a 20 degree rise in temperature and other scenario characteristics. residents reached a plateau in their willingness to pay to prevent 120 degree Fahrenheit mean temperatures as compared to 110 degree Fahrenheit mean temperatures. However, the actual damages to the L.A. basin residents of mean high temperatures of 110 or more degrees Fahrenheit (which would imply occasional extreme heat waves similar in temperature to Death Valley mean highs) would be orders of magnitude more costly, we believe, than a 100 degree Fahrenheit mean high temperature, as such extreme heat would decimate most existing vegetation and threaten the lives of tens of thousands of elderly and other persons vulnerable to heat stroke. For just such reasons, Berk and Shulman strongly caution against taking the dollar values from the survey literally or using them in cost-benefit analyses as they confound several source of value including stewardship and altruism. In addition, some of the climate increases were well above the range of current scientific estimates of greenhouse warming.

The survey was not done in conjunction with atmospheric scientists and climatologists who could provide more realistic climate scenarios or ecologists, public health officials, or others who could help the respondents realise what such warming might mean for trees, birds, or people. Contingent valuation of the hypothetical good is possible when people believe the survey scenario. We present this type of evaluation study to highlight how difficult it is to findacceptable methods to place values on the climatic components of ecosystem services. In this survey case, the background of the respondents as well as their (limited) prior knowledge of the impacts of greenhouse warming played a large role in the survey outcomes. At the same time, however, contingent valuation points out that people are willing to pay to preserve ecosystem services as well as the tremendous need for education to help citizens more realistically value climate and other environmental

The ongoing disturbances of the atmosphere that affect the biogeochemical and physical processes that determine the climate may influence human and natural systems in profound ways. We have attempted to outline a few of the major ecosystem services that are associated with climate and the atmosphere, as well as introduce the challenging task of quantifying, and ultimately monetizing, these services. Current monetized estimates of climate damage by the middle of the 21st century from typical climate change scenarios range from slight economic benefit to a trillion or more dollars lost annually, with most macroeconomic assessments assuming a one to twoper cent annual loss to GWP from climate change. Moreover, the interacting processes and biogeochemical cycles occurring across a wide spectrum of scales lead to synergistic effects that are notusually considered and sometimes not even known when we attempt to disaggregate and value ecosystem services.

The deforestation of the Amazon basin is one example of interacting scales where land use change affecting local and regional climate may also produce a net global residual. Even if the mosaic of regional effects average themselves out globally, there could be residual effects arising from heterogeneous forcing of the climate in areas outside of the tropics (*i.e.* regional high concentrations of sulfate aerosols or tropospheric ozone).

Ecosystems both mediate and respond to the climate system through a variety of physical, biological, and chemical feedback cycles. The uncertainty of resulting synergisms and potential global effects, as exemplified in the Amazon basin, points to the important challenge of defining and understanding the processes that link species and ecosystems with climate. With increasing knowledge, we can better anticipate ecological responses under changing climate scenarios. Meanwhile, humanity continues to perform this potentially trillion dollar unnatural experiment on "Labouratory Earth".

Air, Climate and Weather

The Earth's atmosphere is a blanket of gases approx-imately 350km (218 miles) thick. It is a large and complex system that interacts with the Sun, the land, and the oceans in order to produce both the Earth's weather and climate.

The Earth's Atmosphere has four distinct layers:

1. Thermosphere
2. Mesosphere
3. Stratosphere
4. Troposphere

Thermosphere

The layer of atmosphere most distant from the Earth is thethermosphere, which begins approximately 80km in altitude. It is also the hottest layer; "thermo" being Greek for heat. The temperatures in thethermosphere increase with altitude due to the absorption of intense solar radiation by the limited amount of remaining molecular oxygen. The source of this heat is through bombardment of solar particles carried on the solar wind that do not reach deeper into the atmosphere.

Stratosphere

Thestratosphere extends from approximately 10-12 km to around 50 km above the Earth's surface. The air temperature remains relatively constant up to an altitude of 25 km, then increases gradually having a stabilizing effect on atmospheric conditions. Thestratosphere contains nearly 90per cent of the atmospheric ozone, which plays a major role in regulating temperatures as solar energy is converted to kinetic energy when the ozone molecules absorb ultraviolet radiation, resulting in the heating of thestratosphere.

Troposphere

Thetroposphere is the layer closest to the Earth's surface, containing more than 80per cent of total atmospheric mass composed of 78per cent nitrogen, 21per cent oxygen, other trace gases, water droplets, dust, and other particles. Thetroposphere is where most weather occurs; the circulation of air in intensive vertical movements results in the formation of clouds, while horizontal movements results in wind. Both temperature and water vapour content decrease rapidly as altitude increases within thetroposphere.

Nearly 99per cent of atmospheric water vapour is contained within this layer, which plays a major role in regulating air temperature as it absorbs solar energy and thermal radiation from the planet's surface.

Scientists believe that the Earth's climate is changing and is, in fact, heating up. However, there are considerable differences among the views with regard to the rate of change, the impact on our environment, and what can or should be done about it.

Current environmental concerns are increa-singly focusing on the particles and gases released into the atmosphere as a result of human activity. Frequently, unusual weather events, such as Hurricane Katrina in 2005, are being linked to global warming. But "weather" and "climate" are two very different things, and it is important to understand the difference, as well as the relation, between the two.

In thetroposphere, air rises as it is heated by the sun, falls earthward as it cools, and intermixes with evapourated water from the planet's bodies of water to form clouds and precipitation. The uneven heating of the Earth's surface by the sun, along with Earth's rotation, creates rising (convection), falling (advection), and horizontal air movements (winds). The result of these processes occurring in the form of rain, snow, heat or freezing cold, at a particular place and time, is called weather. The longer term trends in patterns of temperature, rainfall, and other weather indicators over time, usually in blocks of 30 years that can affect the entire Earth, is called climate.

Land and Land Use

Land is the solid part of the Earth's surface, also known as the lithosphere. The surface of the Earth is shaped by a combination of physical processes, including earthquakes and volcanoes, shifts of rocks and sediments, and flows of river and ice. Human activities also shape the land in many ways, including expanding agriculture and population, excavating mineral and forest resources, and changing the flow of rivers with dams and channels.

Land cover is the physical and biological material found on the land surface – which can be either vegetation or man-made structures. Land cover differs from land use, althoughthe way land is used is often either a cause or an effect of its cover type. Land use describes the various ways in which human beings make use of and manage the land and its resources. The way land is used and managed can alter land cover, and can also have an impact on surrounding habitats, the flow of surface waters, and on biogeochemical cycles.

Land cover and land use issues, therefore, can play a significant role in environmental change, including climate change. When land cover is altered, it can result in the destruction of natural carbon sinks that absorb atmospheric carbon dioxide, a heat-trapping gas that contributes to climate change. Deforestation, is a major contributor to observed increases of carbon dioxide in the atmosphere.

There are often competing interests and tradeoffs related to land use that must be taken into consideration. Land use planners and decision makers often struggle with how to meet human needs for food and the utilization of our natural resources while also minimizing impacts and preserving important habitats. These choices are typically made by those who own or control the land; although the choices can also be limited by the physical and biological characteristics of the land, including the climate, soils, and geography.

Land use choices are also limited by institutional factors and, in the United States, can be governed by a mix of local, county, state, and federal laws. Although most of the land. is privately owned, the federal government is the single largest owner of land. Analysis of land use and land cover has been greatly facilitated in the last few decades with the availability of new technologies, including remote sensing. The development of geographic information systems (GIS) has made it possible to combine physical characteristics of the land with information about population, economic data, and other information for further analysis

Food

Human beings began domesticating plants and animals as long as 10,000 years ago. Success i this endeavor led to drastic changes in how and where human beings lived. It changed how humans interacted with each other and how they interacted with the earth. Agriculture led humans from a nomadic existence to one based in permanent and semi-permanent settlements. Human civilization has its roots in the early domestication of plants and animals.

The unprecedented growth in global population that occurred in the twentieth century was made possible by remarkable advances in agriculture and technology. The Green Revolution brought high-yield crops and advanced growing techniques to developing countries and improved nutrition and health for most parts of the world. Continuing population growth will require more acres of land to plant and more water for irrigation, increasing pressure on habitats and natural resources.

Advanced agricultural methods have made it possible to grow more food on fewer acres of land, which permits much land to be returned to forest and other uses. However, there are costs associated with high-yield methods, including their heavy reliance on pesticides and fertilizers. Run-off of fertilizers from agricultural lands has affected the water quality of some lakes, rivers, and bays.

Erosion is, and has been since crops were first planted, a serious environmental problem in the absence of preventative measures. Irrigation, required in many parts of the world, can be a significant drain on water supplies in arid areas and, if improperly managed, can lead to buildup of salt deposits, which may severely degrade soil quality.

One of the most critical challenges of the next few decades is to find ways to increase production of food while minimizing environmental degradation. Considerable research in this and other countries has led to advanced methods to preserve soil and prevent erosion, including conservation tilling and computer-controlled application of pesticides, fertilizers, and water. Pesticide use is controversial. Other methods, especially new types of genetic modification, or "genetic engineering" are more controversial.

To reduce reliance on pesticides, methods such as biocontrol (using other species to control pests) and biotechnology to create pest-resistant crops are being tested. Biocontrol using wild-type (naturally occurring) species is usualy accepted, especially by organic producers. Naturally occurring mutants are accepted as well. More controversial is the use of genetically modified biocontrol agents changed by DNA deletion or insertion of new traits. Concerns about the potential impacts of these methods have been raised.

Environment and Society

Throughout history humans have both affected, and been affected by, the natural world. While a good deal has been lost due to human actions, much of what is valued about the environment has been preserved and protected through human action. While many uncertainties remain, there is a realization that environmental problems are becoming more and more complex, especially as issues arise on a more global level, such as that of atmospheric pollution or global warming.

Interactions between human society and the environment are constantly changing. The environment, while highly valued by most, is used and altered by a wide variety of people with many different interests and values. Difficulties remain on how best to ensure the protection of our environment and natural resources. There will always be tradeoffs and, many times, unanticipated or unintended consequences. Hwever, a well-managed environment can provide goods and services that are both essential for our well being as well as for continued economic prosperity.

Chapter

4

Bio Nutrients

The main flow of energy that helps shape conditions on Earth's surface comes from space; and when the energy's work here is done, to space it returns. With respect to energy, then, Earth is an open system. With respect to its chemical endowment, however, Earth is a closed system. That is, the amounts of carbon, hydrogen, oxygen, iron, gold, and other elements in the planet-atmosphere system do not change with time; the chemical arrangement and physical distribution of these elements can and do vary, but essentially nothing enters and nothing leaves the system.

The elements in this closed system that are essential to life are called *nutrients*. They can be divided into three categories: the 4 main chemical building blocks of living matter (carbon, oxygen, hydrogen, nitrogen); 7 *macronutrients*, of which smaller but still signficant quantities are required for life; and 13 *micronutrients*, or trace elements, of which tiny quantities perform essential functions. The suitability of any terrestrial or aquatic environment for the support of life depends on the availability of nutrients in appropriate forms and quantities. The processes that govern this availbility (or lack of it) are known collectively as *ntrient cycles*, becauseof the way the individual basic stocks of physical material move cyclically through the living and nonliving parts of the physical world.

These cycles function in support of life not merely by making nutrients continuously available—in other words, by maintaining the fertility of the environment -but also by limiting the accumulation of material in quantities, forms, and plaes in which it would damage organisms. It is important to recognize in this connection that the same elements and compounds that serve as nutrients for some organisms in some concentrations are often toxic for other organisms, or even for the same organisms at higher concentrations.

For example, molecular oxygen (O_2) is toxic to anaerobic organisms and, at high enough concentrations, even to mammals. Ammonia (NH_3) is an important source of nutrient nitrogen for many plants but is toxic to people. Hydrogen sulfide (H_2S) is a nutrient for certain types of bacteria but is extremely toxic to mammals.

Dynamics of Nutrient Cycling

The steps that make up all nutrient cycles involve two basic processes: physical transport and chemical transformation.

They are:

- The hydrologic cycle, including evaporation and precipitation as well as the flow of streams, groundwater, and ice;
- Winds;
- Ocean currents;
- Geologic movement, especially the upward and downward motions at the boundaries of tectonic plates and geologic uplifting on continents.

A fifth agent of physical transport that is important in a few circumstances is the movement of organisms. For example, fish-eating birds that deposit their excrement on land provide a significant pathway by which nitrogen and phosphorus are transferred from the sea to the land. Fish such as salmon, that feed mostly in the ocean but migrate up freshwater rivers to spawn and die, perform a similar function, as do the ocean-caught fish consumed by continent-dwelling human beings.

How the various mechanisms of transport link the principal "compartments" of the physical world. Obviously, the ease (and, hence, the rate) with which a given substance can move through these pathways depends on its physical state (solid, liquid, gas), on other physical properties (solubility, volatility), and on the possibility of chemical transformations that can change those properties. Such chemical transformations take place in living organisms and in the nonliving environment.

They are considered in more detail below. Each of the compartments indicated often is divided into subcompa-rtments for the purpose of tracing nutrient flows. Thus, for example, it is customary to consider three subcompartments of material in both the soil and the oceans: living organic matter, dead organic matter, and the inorganic medium itself.

These subcompartments can be further subdivided—for example, organic matter, into plants, animals, and microorganisms; and seawater, into the well-mixed surface layer and the stratified deeper layers. Finally, within each subdivision of each subcompartment, the magnitudes of the stocks, or pools, of the important elements and compounds are of interest, as are the magnitudes of the flows into and out of the pools (where flows can mean physical movement or chemical transformation or a combination of the two).

For any pool that is in steady state—that is, for which inflows balance outflows so the size of the pool does not change with time—an average residence time for a compound can be calculated as the ratio of the size of the pool to the sum of the inflows or outflows.

Examination of the residence times for important pools provides quick insight into the dynamics of nutrient cycling, and, indirectly, into the probable vulnerability of specific components of nutrient cycles to disruption by human intervention.

For example, a long residence time means the annual flows are small compared to the pool, which means in turn that unless human disruptions cause a very large alteration in inflow or outflow they will affect the size of the pool only very slowly. A short residence time, by contrast, means that significant changes in inflow or outflow can affect the size of the pool very quickly.

Chemistry of Nutrient Cycles

Two kinds of chemistry play crucial roles in the principal global nutrient cycles: the chemistry of the sedimentary cycle, associated with the very slow transformation of very large pools of material, and the chemistry of life, associated with the more rapid transformation of much smaller pools.

Both kinds of chemistry depend very heavily on the very special roles ofwater as a source of hydrogen (H^+) and hydroxyl (OH^-) ions for many chemical reactions, as a medium for holding other chemical compounds in solution where they can react, and as a vehicle for physical transport of compounds from place to place.

Cycles of The Principal Nutrients

The cycles of many nutrients are tightly linked chemically and biologically. This is so much the case for the cycles of carbon, oxygen, and hydrogen that we consider them together. The cycles of nitrogen, phosphorus, and sulfur, on the other hand, can usefully be examined separately.

Carbon-oxygen-hydrogen

It is appropriate to begin our closer examination of the carbon-oxygen-hydrogen cycles with their intersection in the green plant, the basis of almost all life on Earth.

Carbon and Energy in Living Plants

Most living plants not only fix (store) energy by means of photosynthesis but they also use some of that stored energy to drive their own life-sustaining internal processes by means of respiration.

The total rate at which a plant or a plant community stores solar energy by means of photosynthesis is called *gross photosynthesis or gross primary production* (GPP). The figure obtained by subtracting from gross photosynthesis the amount of energy used by the plants in respiration (R) is called *net photosynthesis or net primary production* (NPP). In abbreviated form, then, the relation is GPP-R = NPP. Several interchangeable units are used in the biological literature for the measurement of these quantities: calories or kilocalories per unit of time, watts or kilowatts (different units for measuring the same thing—energy flow), grams of carbon fixed into organic compounds per unit of time, moles of carbon fixed per unit of time, or grams of dry organic matter produced per unit of time.

A wide range of estimates of global net primary production and the associated standing crop of living plants (*phytomass*) have been published. One recent review indicates that the estimates of net primary production are converging on a value around 160 billion (1.60 x 10) metric tons per year for the globe. Global phytomass was estimated in the same review as about ten times larger 1840 billion metric tons, dry weight and gross photosynthesis was estimated to be about twice as large as net production (that is, half of gross photosynthesis is consumed in plant respiration and half remains as net production). The estimates for phytomass and net primary production, separated into terrestrial and oceanic components. The roughness of these estimates deserves emphasis. The range of recent estimates of land phytomass 30 per cent larger; the range of published values of ocean phytomass is from twenty times smaller than the value to three times larger. The factor-of-60 range of informed opinion with respect to oceanic phytomass is especially remarkable.

The range in recent published estimates of oceanic net primary production is only ± 10 per cent from the value, and the range in terrestrial net production is from 15 per cent less to 60 per cent more than the value in the table. The fraction of gross primary production consumed in plant respiration is known to vary between 20 per cent and 75 per cent, depending on type of plant and region, but the global average is unlikely to differ much from 50 per cent.

Energy and Material Flow in the Carbon Cycle

For the purpose of reconciling estimates of annual production presented in grams of carbon, grams of dry organic matter, and kilocalories, one widely used convention assumes that 10 kilocalories of energy are stored per gram of fixed carbon and that dry organic matter contains 45 per cent fixed carbon by weight. (The actual range of values for different kinds of dry organic matter extends roughly ± 15 per cent from these nominal figures.) Thus, measurements in grams or metric tons (1 MT = 106 g) of carbon can be converted to grams or metric tons of dry organic matter by multiplying by 2.2, and 1 gram of dry organic matter can be taken to represent 4.5 kilocalories of stored energy.

Useful conversions following directly from this convention are that I gram of dry organic matter per day corresponds to an average energy flow of about 0.2 watts; 1 gram of carbon per day, to about 0.5 watts; and one metric ton of dry organic matter per year, to about 600 watts. Also, 1 metric ton of dry organic matter contains 37,500 moles of carbon. Although we shall use the convention just described, the reader may encounter others elsewhere. If the dry organic matter were pure carbohydrate $(CH_2O)n$ which stores 112 kilocalories per mole of carbon, then the appropriate conversions would be: 3.7 kilocalories per gram of dry organic matter, 9.3 kilocalories per gram of carbon, 0.4 gram carbon per gram of dry organic matter [Actually, CH_2O is a good approximation of the average composition of all living matter because so much of the world's biomass is cellulose tied up in woody plants, and cellulose has nearly this formula.

However, most of the plant material *produced* each year is not cellulose—remember, a big pool does not necessarily mean a big flow. The annual production of organic matter contains a higher proportion of fixed carbon and higher energy content per gram than does the cellulose-dominated, slow-toturn-over standing crop.] The rounded-off value of 4 kilocalories per gram of dry organic matter is in fairly widespread use in conjunction with the figure of 10 kilocalories per gram of fixed carbon, itself a rough average among several kinds of plant matter.

Taking the range of estimates into account, one can state several conclusions with reasonable assurance:

- The annual NPP of the oceans is only one-third to one-half that of the continents, even though the oceans cover more than 70 per cent of Earth's surface;
- The pool, or standing crop, of living plant material on land is at least several hundred times that in the oceans;
- The average turnover time (life expectancy) of oceanic plants must be in the range of days to weeks, in order for so small a phytomass to be associated with so large an annual production;
- The average turnover time of terrestrial living plants, by contrast, is in the range of a decade or two;
- Oceanic and terrestrial phytomasses are both so poorly known that man-induced or natural changes large enough to be very important in the long term could easily escape notice for some time. In the latter regard, although there are good reasons to suppose that the activities of civilization have diminished the terrestrial phytomass significantly, some experts consider this assertion to be debatable.

Notwithstanding the uncertainties in the data, it is instructive to examine the large regional variations in plant productivity that are concealed in the global totals. Of particular interest are the facts that tropical forests account for more than 40 per cent of terrestrial NPP (although they cover only about 16 per cent of the land area) and that up- welling regions, continental shelves, reefs, and estuaries yield almost 25 per cent of oceanic NPP while accounting for less than 9 per cent of the ocean's area.

The rate at which energy is being stored in net primary production is about 100 million megawatts, roughly two-thirds on land and one-third in the ocean. The rate of gross primary production can be assumed to be about 200 million megawatts. The net primary production is roughly twelve times the rate at which civilization used commercial energy in 1975, where commercial energy refers to energy that is sold (for example, fossil fuels, hydropower, and nuclear power, but excluding food) as opposed to energy that is gathered by the user (for example, most wood, dung). Very little of the solar energy that reaches Earth's surface is captured by photosynthesis.

The net primary production is about 0.1 per cent of the incident solar energy at the surface-0.25 per cent of what strikes the land and 0.05 per cent of what strikes the ocean. Some of the reasons these percentages are so small are obvious. Plants do not by any means cover all of Earth's surface; in many areas of the world the growing season lasts only part of the year; and in some circumstances, as already noted, plant growth is limited by lack of one or more critical nutrients.

Under favourable conditions, on the other hand—as in a well watered and fertilized cornfield or a fertile swamp—NPP may reach 2 per cent or so of incident solar energy averaged over the growing season and as much as 6 per cent or 7 per cent of incident solar energy on the most favourable days. One additional reason for this is that only about one-fourth the solar energy reaching the ground is in the part of the wavelength range that stimulates photos-ynthesis (blue and red light).

The other big reason is the large amount of water that must be evaporated through the leaves of plants to maintain the nutrient-bearing flow of water through them. Evaporating this water typically takes 40 per cent or more of the energy that falls on a plant, and a much higher percentage of what is actually absorbed by the plant. (For production of every ton of wet weight of many crops—of which about one-fourth is dry organic matter around 100 tons of water are evaporated.

This means 400 grams of water are evaporated, requiring about 0.6 kilocalories per gram evaporated, for every gram or 4.5 kilocalories of dry organic matter produced. The ratio in such a case is about 50 kilocalories for evaporation per kilocalorie of net primary production.

Oxygen Balance: Metabolism by Animals and Microbes

The overall effects of the net production of plants on the cycles of carbon, oxygen, and hydrogen are as follows:

- More oxygen, O_2, is added to the environment than is removed from it;
- Less carbon dioxide, CO_2, and water, H_2O, are added to the environment than are removed from it;

- The CO_2 and H_2O subtracted from the environmental pools show up as additions to the pool of stored organic matter, CH_2O, and to the atmospheric oxygen pool. Quantitatively, for every metric ton of dry organic matter produced, there is a net loss from the environment of 0.6 metric tons of water and about 1.5 metric tons of CO_2, and a net gain of about 1.1 metric tons of oxygen.

Why, then, is oxygen not building up continuously in the atmosphere, while water and carbon dioxide are depleted? Because plant matter is not accumulating, but rather is approximately in steady state. It is being broken down into carbon dioxide and water again by the metabolism of animals and microbes, using up molecular oxygen just as rapidly as new plant material and oxygen are being produced.

Consider the possible fates of plant carbohydrate produced as part of net primary production. Plants either are consumed by herbivores, or they die and add their carbohydrate to the pool of dead organic matter. Of the part consumed by herbivores, some is metabolized and thereby returned to the environmental pool of CO_2 and H_2O, some is added as excrement to the pool of dead organic matter, and some is incorporated into herbivore tissue, either to be eaten by carnivores or to be added at the herbivore's death (by causes other than being devoured) directly to the pool of dead organic matter. The same fates await the carbohydrate that reaches the carnivores. The cycle is closed by the action of decomposers—both animals and microbes (bacteria)—which by metabolizing the fixed carbon in dead organic matter return CO and H O to the environmental pools.

Chapter

5

Enzymes: Mechanisms and Control

Enzymes are proteins that catalyse (*i.e.*, increase or decrease the rates of) chemical reactions. In enzymatic reactions, the molecules at the beginning of the process are called substrates, and they are converted into different molecules, called the products. Almost all processes in a biological cell need enzymes to occur at significant rates. Since enzymes are selective for their substrates and speed up only a few reactions from among many possibilities, the set of enzymes made in a cell determines which metabolic pathways occur in that cell.

Like all catalysts, enzymes work by lowering the activation energy ($E_a^‡$) for a reaction, thus dramatically increasing the rate of the reaction. As a result, products are formed faster and reactions reach their equilibrium state more rapidly. Most enzyme reaction rates are millions of times faster than those of comparable un-catalysed reactions.

As with all catalysts, enzymes are not consumed by the reactions they catalyse, nor do they alter the equilibrium of these reactions. However, enzymes do differ from most other catalysts by being much more specific. Enzymes are known to catalyse about 4,000 biochemical reactions. A few RNA molecules called ribozymes also catalyse reactions, with an important example being some parts of the ribosome. Synthetic molecules called artificial enzymes also display enzyme-like catalysis. Enzyme activity can be affected by other molecules. Inhibitors are molecules that decrease enzyme activity; activators are molecules that increase activity.

Many drugs and poisons are enzyme inhibitors. Activity is also affected by temperature, chemical environment (*e.g.*, pH), and the concentration of substrate. Some enzymes are used commercially, for example, in the synthesis of antibiotics. In addition, some household products use enzymes to speed up biochemical reactions (*e.g.*, enzymes in biological washing powders break down protein or fat stains on clothes; enzymes in meat tenderizers break down proteins, making the meat easier to chew).

Structures and Mechanisms

Enzymes are generally globular proteins and range from just 62 amino acid residues in size, for the monomer of 4-oxalocrotonate tautomerase, to over 2,500 residues in the animal fatty acid synthase. A small number of RNA-based biological catalysts exist, with the most common being the ribosome; these are referred to as either RNA-enzymes or ribozymes. The activities of enzymes are determined by their three-dimensional structure. However, although structure does determine function, predicting a novel enzyme's activity just from its structure is a very difficult problem that has not yet been solved.

Most enzymes are much larger than the substrates they act on, and only a small portion of the enzyme (around 3–4 amino acids) is directly involved in catalysis. The region that contains these catalytic residues, binds the substrate, and then carries out the reaction is known as the active site.

Enzymes can also contain sites that bind cofactors, which are needed for catalysis. Some enzymes also have binding sites for small molecules, which are often direct or indirect products or substrates of the reaction catalysed. This binding can serve to increase or decrease the enzyme's activity, providing a means for feedback regulation.

Like all proteins, enzymes are long, linear chains of amino acids that fold to produce a three-dimensional product. Each unique amino acid sequence produces a specific structure, which has unique properties. Individual protein chains may sometimes group together to form a protein complex. Most enzymes can be denatured—that is, unfolded and inactivated—by heating or chemical denaturants, which disrupt the three-dimensional structure of the protein.

Depending on the enzyme, denaturation may be reversible or irreversible. Structures of enzymes in complex with substrates or substrate analogs during a reaction may be obtained using Time resolved crystallography methods.

Specificity

Enzymes are usually very specific as to which reactions they catalyse and the substrates that are involved in these reactions. Complementary shape, charge and hydrophilic/hydrophobic characteristics of enzymes and substrates are responsible for this specificity. Enzymes can also show impressive levels of stereospecificity, regioselectivity and chemoselectivity.

Some of the enzymes showing the highest specificity and accuracy are involved in the copying and expression of the genome. These enzymes have "proof-reading" mechanisms. Here, an enzyme such as DNA polymerase catalyzes a reaction in a first step and then checks that the product is correct in a second step. This two-step process results in average error rates of less than 1 error in 100 million reactions in high-fidelity mammalian polymerases. Similar proofreading mechanisms are also found in RNA polymerase, aminoacyl tRNA synthetases and ribosomes.

Some enzymes that produce secondary metabolites are described as promiscuous, as they can act on a relatively broad range of different substrates. It has been suggested that this broad substrate specificity is important for the evolution of new biosynthetic pathways.

"Lock and Key" Model

Enzymes are very specific, and it was suggested by the Nobel laureate organic chemist Emil Fischer in 1894 that this was because both the enzyme and the substrate possess specific complementary geometric shapes that fit exactly into one another. This is often referred to as "the lock and key" model. However, while this model explains enzyme specificity, it fails to explain the stabilization of the transition state that enzymes achieve.

In 1958, Daniel Koshland suggested a modification to the lock and key model: since enzymes are rather flexible structures, the active site is continually reshaped by interactions with the substrate as the substrate interacts with the enzyme. As a result, the substrate does not simply bind to a rigid active site; the amino acid side chains which make up the active site are molded into the precise positions that enable the enzyme to perform its catalytic function. In some cases, such as glycosidases, the substrate molecule also changes shape slightly as it enters the active site. The active site continues to change until the substrate is completely bound, at which point the final shape and charge is determined. Induced fit may enhance the fidelity of molecular recognition in the presence of competition and noise via the conformational proofreading mechanism.

Mechanisms

Enzymes can act in several ways, all of which lower $\Delta G^{\ddagger}$:

- Lowering the activation energy by creating an environment in which the transition state is stabilized (e.g. straining the shape of a substrate—by binding the transition-state conformation of the substrate/product molecules, the enzyme distorts the bound substrate(s) into their transition state form, thereby reducing the amount of energy required to complete the transition).
- Lowering the energy of the transition state, but without distorting the substrate, by creating an environment with the opposite charge distribution to that of the transition state.
- Providing an alternative pathway. For example, temporarily reacting with the substrate to form an intermediate ES complex, which would be impossible in the absence of the enzyme.
- Reducing the reaction entropy change by bringing substrates together in the correct orientation to react. Considering $\Delta H^{\ddagger}$ alone overlooks this effect.
- Increases in temperatures speed up reactions. Thus, temperature increases help the enzyme function and develop the end product even faster. However, if heated too much, the enzyme's shape deteriorates and only when the temperature comes back to normal does the enzyme regain its shape. Some enzymes like thermolabile enzymes work best at low temperatures.

Interestingly, this entropic effect involves destabilization of the ground state, and its contribution to catalysis is relatively small.

Transition State Stabilization

The understanding of the origin of the reduction of $\Delta G^{\ddagger}$ requires one to find out how the enzymes can stabilize its transition state more than the transition state of the uncatalyzed reaction. Apparently, the most effective way for reaching large stabilization is the use of electrostatic effects, in particular, by having a relatively fixed polar environment that is oriented toward the charge distribution of the transition state. Such an environment does not exist in the uncatalyzed reaction in water.

Dynamics and Function

The internal dynamics of enzymes is linked to their mechanism of catalysis. Internal dynamics are the movement of parts of the enzyme's structure, such as individual amino acid residues, a group of amino acids, or even an entire protein domain. These movements occur at various time-scales ranging from femtoseconds to seconds. Networks of protein residues throughout an enzyme's structure can contribute to catalysis through dynamic motions.

Protein motions are vital to many enzymes, but whether small and fast vibrations, or larger and slower conformational movements are more important depends on the type of reaction involved. However, although these movements are important in binding and releasing substrates and products, it is not clear if protein movements help to accelerate the chemical steps in enzymatic reactions. These new insights also have implications in understanding allosteric effects and developing new drugs.

Allosteric Modulation

Allosteric sites are sites on the enzyme that bind to molecules in the cellular environment. The sites form weak, noncovalent bonds with these molecules, causing a change in the conformation of the enzyme. This change in conformation translates to the active site, which then affects the reaction rate of the enzyme. Allosteric interactions can both inhibit and activate enzymes and are a common way that enzymes are controlled in the body.

Cofactors and Coenzymes

Cofactors

Some enzymes do not need any additional components to show full activity. However, others require non-protein molecules called cofactors to be bound for activity. Cofactors can be either inorganic (*e.g.*, metal ions and iron-sulfur clusters) or organic compounds (e.g., flavin and heme). Organic cofactors can be either prosthetic groups, which are tightly bound to an enzyme, or coenzymes, which are released from the enzyme's active site during the reaction. Coenzymes include NADH,

NADPH and adenosine triphosphate. These molecules transfer chemical groups between enzymes. An example of an enzyme that contains a cofactor is carbonic anhydrase, and is shown in the ribbon diagram above with a zinc cofactor bound as part of its active site. These tightly bound molecules are usually found in the active site and are involved in catalysis. For example, flavin and heme cofactors are often involved in redox reactions.

Enzymes that require a cofactor but do not have one bound are called *apoenzymes* or *apoproteins*. An apoenzyme together with its cofactor(s) is called a *holoenzyme* (this is the active form). Most cofactors are not covalently attached to an enzyme, but are very tightly bound.

However, organic prosthetic groups can be covalently bound (*e.g.*, thiamine pyrophosphate in the enzyme pyruvate dehydrogenase). The term "holoenzyme" can also be applied to enzymes that contain multiple protein subunits, such as the DNA polymerases; here the holoenzyme is the complete complex containing all the subunits needed for activity.

Coenzymes

Coenzymes are small organic molecules that transport chemical groups from one enzyme to another. Some of these chemicals such as riboflavin, thiamine and folic acid are vitamins (compounds which cannot be synthesized by the body and must be acquired from the diet).

The chemical groups carried include the hydride ion (H^-) carried by NAD or $NADP^+$, the phosphate group carried by adenosine triphosphate, the acetyl group carried by coenzyme A, formyl, methenyl or methyl groups carried by folic acid and the methyl group carried by S-adenosylmethionine.

Since coenzymes are chemically changed as a consequence of enzyme action, it is useful to consider coenzymes to be a special class of substrates, or second substrates, which are common to many different enzymes. For example, about 700 enzymes are known to use the coenzyme NADH.

Coenzymes are usually continuously regenerated and their concentrations maintained at a steady level inside the cell: for example, NADPH is regenerated through the pentose phosphate pathway and *S*-adenosylmethionine by methionine adenosyltransferase. This continuous regeneration means that even small amounts of coenzymes are used very intensively. For example, the human body turns over its own weight in ATP each day.

Kinetics

Enzyme kinetics is the investigation of how enzymes bind substrates and turn them into products. The rate data used in kinetic analyses are obtained from enzyme assays. In 1902 Victor Henri proposed a quantitative theory of enzyme kinetics, but his experimental data were not useful because the significance of the hydrogen ion concentration was not yet appreciated.

After Peter Lauritz Sørensen had defined the logarithmic pH-scale and introduced the concept of buffering in 1909 the German chemist Leonor Michaelis and his Canadian postdoc Maud Leonora Menten repeated Henri's experiments and confirmed his equation which is referred to as Henri-Michaelis-Menten kinetics (sometimes also Michaelis-Menten kinetics). Their work was further developed by G. E. Briggs and J. B. S. Haldane, who derived kinetic equations that are still widely used today.

The major contribution of Henri was to think of enzyme reactions in two stages. In the first, the substrate binds reversibly to the enzyme, forming the enzyme-substrate complex. This is sometimes called the Michaelis complex. The enzyme then catalyzes the chemical step in the reaction and releases the product.

Enzymes can catalyse up to several million reactions per second. For example, the uncatalyzed decarboxylation of orotidine 5'-monophosphate has a half life of 78 million years. However, when the enzyme orotidine 5'-phosphate decarboxylase is added, the same process takes just 25 milliseconds. Enzyme rates depend on solution conditions and substrate concentration.

Conditions that denature the protein abolish enzyme activity, such as high temperatures, extremes of pH or high salt concentrations, while raising substrate concentration tends to increase activity. To find the maximum speed of an enzymatic reaction, the substrate concentration is increased until a constant rate of product formation is seen. This is shown in the saturation curve on the right.

Saturation happens because, as substrate concentration increases, more and more of the free enzyme is converted into the substrate-bound ES form. At the maximum velocity (V_{max}) of the enzyme, all the enzyme active sites are bound to substrate, and the amount of ES complex is the same as the total amount of enzyme. However, V_{max} is only one kinetic constant of enzymes. The amount of substrate needed to achieve a given rate of reaction is also important. This is given by the Michaelis-Menten constant (K_m), which is the substrate concentration required for an enzyme to reach one-half its maximum velocity. Each enzyme has a characteristic K_m for a given substrate, and this can show how tight the binding of the substrate is to the enzyme. Another useful constant is k_{cat}, which is the number of substrate molecules handled by one active site per second.

The efficiency of an enzyme can be expressed in terms of k_{cat}/K_m. This is also called the specificity constant and incorporates the rate constants for all steps in the reaction. Because the specificity constant reflects both affinity and catalytic ability, it is useful for comparing different enzymes against each other, or the same enzyme with different substrates.

The theoretical maximum for the specificity constant is called the diffusion limit and is about 10^8 to 10^9 (M^{-1} s^{-1}). At this point every collision of the enzyme with its substrate will result in catalysis, and the rate of product formation is not limited by the reaction rate but by the diffusion rate.

Enzymes with this property are called *catalytically perfect* or *kinetically perfect*. Example of such enzymes are triose-phosphate isomerase, carbonic anhydrase, acetylcholinesterase, catalase, fumarase, β-lactamase, and superoxide dismutase.

Michaelis-Menten kinetics relies on the law of mass action, which is derived from the assumptions of free diffusion and thermodynamically driven random collision. However, many biochemical or cellular processes deviate significantly from these conditions, because of macromolecular crowding, phase-separation of the enzyme/substrate/product, or one or two-dimensional molecular movement. In these situations, a fractal Michaelis-Menten kinetics may be applied.

Some enzymes operate with kinetics which are faster than diffusion rates, which would seem to be impossible. Several mechanisms have been invoked to explain this phenomenon. Some proteins are believed to accelerate catalysis by drawing their substrate in and pre-orienting them by using dipolar electric fields. Other models invoke a quantum-mechanical tunnelling explanation, whereby a proton or an electron can tunnel through activation barriers, although for proton tunnelling this model remains somewhat controversial.

Quantum tunnelling for protons has been observed in tryptamine. This suggests that enzyme catalysis may be more accurately characterized as "through the barrier" rather than the traditional model, which requires substrates to go "over" a lowered energy barrier.

Inhibition

Enzyme reaction rates can be decreased by various types of enzyme inhibitors.

Competitive Inhibition

In competitive inhibition, the inhibitor and substrate compete for the enzyme (i.e., they can not bind at the same time). Often competitive inhibitors strongly resemble the real substrate of the enzyme. For example, methotrexate is a competitive inhibitor of the enzyme dihydrofolate reductase, which catalyzes the reduction of dihydrofolate to tetrahydrofolate.

The similarity between the structures of folic acid and this drug. Note that binding of the inhibitor need *not* be to the substrate binding site (as frequently stated), if binding of the inhibitor changes the conformation of the enzyme to prevent substrate binding and *vice versa*. In competitive inhibition the maximal velocity of the reaction is not changed, but higher substrate concentrations are required to reach a given velocity, increasing the apparent K_m.

Uncompetitive Inhibition

In uncompetitive inhibition the inhibitor can not bind to the free enzyme, but only to the ES-complex. The EIS-complex thus formed is enzymatically inactive. This type of inhibition is rare, but may occur in multimeric enzymes.

Non-competitive Inhibition

Non-competitive inhibitors can bind to the enzyme at the same time as the substrate, i.e. they *never* bind to the active site. Both the EI and EIS complexes are enzymatically inactive. Because the inhibitor can not be driven from the enzyme by higher substrate concentration (in contrast to competitive inhibition), the apparent V_{max} changes. But because the substrate can still bind to the enzyme, the K_m stays the same.

Mixed Inhibition

This type of inhibition resembles the non-competitive, except that the EIS-complex has residual enzymatic activity.

In many organisms inhibitors may act as part of a feedback mechanism. If an enzyme produces too much of one substance in the organism, that substance may act as an inhibitor for the enzyme at the beginning of the pathway that produces it, causing production of the substance to slow down or stop when there is sufficient amount. This is a form of negative feedback. Enzymes which are subject to this form of regulation are often multimeric and have allosteric binding sites for regulatory substances.

Their substrate/velocity plots are not hyperbolar, but sigmoidal (S-shaped). Irreversible inhibitors react with the enzyme and form a covalent adduct with the protein. The inactivation is irreversible. These compounds include eflornithine a drug used to treat the parasitic disease sleeping sickness.

Penicillin and Aspirin also act in this manner. With these drugs, the compound is bound in the active site and the enzyme then converts the inhibitor into an activated form that reacts irreversibly with one or more amino acid residues.

Uses of Inhibitors

Since inhibitors modulate the function of enzymes they are often used as drugs. An common example of an inhibitor that is used as a drug is aspirin, which inhibits the COX-1 and COX-2 enzymes that produce the inflammation messenger prostaglandin, thus suppressing pain and inflammation.

However, other enzyme inhibitors are poisons. For example, the poison cyanide is an irreversible enzyme inhibitor that combines with the copper and iron in the active site of the enzyme cytochrome c oxidase and blocks cellular respiration.

Biological Function

Enzymes serve a wide variety of functions inside living organisms. They are indispensable for signal transduction and cell regulation, often via kinases and phosphatases. They also generate movement, with myosin hydrolysing ATP to generate muscle contraction and also moving cargo around the cell as part of the cytoskeleton.

Other ATPases in the cell membrane are ion pumps involved in active transport. Enzymes are also involved in more exotic functions, such as luciferase generating light in fireflies. Viruses can also contain enzymes for infecting cells, such as the HIV integrase and reverse transcriptase, or for viral release from cells, like the influenza virus neuraminidase.

An important function of enzymes is in the digestive systems of animals. Enzymes such as amylases and proteases break down large molecules (starch or proteins, respectively) into smaller ones, so they can be absorbed by the intestines. Starch molecules, for example, are too large to be absorbed from the intestine, but enzymes hydrolyse the starch chains into smaller molecules such as maltose and eventually glucose, which can then be absorbed.

Different enzymes digest different food substances. In ruminants which have herbivorous diets, microorganisms in the gut produce another enzyme, cellulase to break down the cellulose cell walls of plant fiber.

Several enzymes can work together in a specific order, creating metabolic pathways. In a metabolic pathway, one enzyme takes the product of another enzyme as a substrate. After the catalytic reaction, the product is then passed on to another enzyme. Sometimes more than one enzyme can catalyse the same reaction in parallel, this can allow more complex regulation: with for example a low constant activity being provided by one enzyme but an inducible high activity from a second enzyme.

Enzymes determine what steps occur in these pathways. Without enzymes, metabolism would neither progress through the same steps, nor be fast enough to serve the needs of the cell. Indeed, a metabolic pathway such as glycolysis could not exist independently of enzymes. Glucose, for example, can react directly with ATP to become phosphorylated at one or more of its carbons.

In the absence of enzymes, this occurs so slowly as to be insignificant. However, if hexokinase is added, these slow reactions continue to take place except that phosphorylation at carbon 6 occurs so rapidly that if the mixture is tested a short time later, glucose-6-phosphate is found to be the only significant product. Consequently, the network of metabolic pathways within each cell depends on the set of functional enzymes that are present.

Control of Activity

There are five main ways that enzyme activity is controlled in the cell.

1. Enzyme production (transcription and translation of enzyme genes) can be enhanced or diminished by a cell in response to changes in the cell's environment. This form of gene regulation is called enzyme induction and inhibition. For example, bacteria may become resistant to antibiotics such as penicillin because enzymes called beta-lactamases are induced that hydrolyse the crucial beta-lactam ring within the penicillin molecule.

Another example are enzymes in the liver called cytochrome P450 oxidases, which are important in drug metabolism. Induction or inhibition of these enzymes can cause drug interactions.

2. Enzymes can be compartmentalized, with different metabolic pathways occurring in different cellular compartments. For example, fatty acids are synthesized by one set of enzymes in the cytosol, endoplasmic reticulum and the Golgi apparatus and used by a different set of enzymes as a source of energy in the mitochondrion, through b-oxidation.
3. Enzymes can be regulated by inhibitors and activators. For example, the end product(s) of a metabolic pathway are often inhibitors for one of the first enzymes of the pathway (usually the first irreversible step, called *committed step*), thus regulating the amount of end product made by the pathways. Such a regulatory mechanism is called a negative feedback mechanism, because the amount of the end product produced is regulated by its own concentration. Negative feedback mechanism can effectively adjust the rate of synthesis of intermediate metabolites according to the demands of the cells. This helps allocate materials and energy economically, and prevents the manufacture of excess end products. The control of enzymatic action helps to maintain a stable internal environment in living organisms.
4. Enzymes can be regulated through post-translational modification. This can include phosphorylation, myristoylation and glycosylation. For example, in the response to insulin, the phosphorylation of multiple enzymes, including glycogen synthase, helps control the synthesis or degradation of glycogen and allows the cell to respond to changes in blood sugar. Another example of post-translational modification is the cleavage of the polypeptide chain. Chymotrypsin, a digestive protease, is produced in inactive form as chymotrypsinogen in the pancreas and transported in this form to the stomach where it is activated. This stops the enzyme from digesting the pancreas or other tissues before it enters the gut. This type of inactive precursor to an enzyme is known as a zymogen.
5. Some enzymes may become activated when localized to a different environment (e.g. from a reducing (cytoplasm) to an oxidizing (periplasm) environment, high pH to low pH etc.).

For example, hemagglutinin in the influenza virus is activated by a conformational change caused by the acidic conditions, these occur when it is taken up inside its host cell and enters the lysosome.

Involvement in Disease

Since the tight control of enzyme activity is essential for homeostasis, any malfunction (mutation, overproduction, underproduction or deletion) of a single critical enzyme can lead to a genetic disease. The importance of enzymes is shown by the fact that a lethal illness can be caused by the malfunction of just one type of enzyme out of the thousands of types present in our bodies. One example is the most common type of phenylketonuria.

A mutation of a single amino acid in the enzyme phenylalanine hydroxylase, which catalyzes the first step in the degradation of phenylalanine, results in build-up of phenylalanine and related products. This can lead to mental retardation if the disease is untreated. Another example is when germline mutations in genes coding for DNA repair enzymes cause hereditary cancer syndromes such as xeroderma pigmentosum. Defects in these enzymes cause cancer since the body is less able to repair mutations in the genome. This causes a slow accumulation of mutations and results in the development of many types of cancer in the sufferer.

Naming Conventions

An enzyme's name is often derived from its substrate or the chemical reaction it catalyzes, with the word ending in-*ase*. Examples are lactase, alcohol dehydrogenase and DNA polymerase. This may result in different enzymes, called isozymes, with the same function having the same basic name. Isoenzymes have a different amino acid sequence and might be distinguished by their optimal pH, kinetic properties or immunologically. Furthermore, the normal physiological reaction an enzyme catalyzes may not be the same as under artificial conditions. This can result in the same enzyme being identified with two different names. *E.g.* Glucose isomerase, used industrially to convert glucose into the sweetener fructose, is a xylose isomerase *in vivo*.

The International Union of Biochemistry and Molecular Biology have developed a nomenclature for enzymes, the EC numbers; each enzyme is described by a sequence of four numbers preceded by "EC". The first number broadly classifies the enzyme based on its mechanism.

The top-level classification is:

- EC 1 *Oxidoreductases*: catalyse oxidation/reduction reactions
- EC 2 *Transferases*: transfer a functional group (*e.g.* a methyl or phosphate group)
- EC 3 *Hydrolases*: catalyse the hydrolysis of various bonds
- EC 4 *Lyases*: cleave various bonds by means other than hydrolysis and oxidation
- EC 5 *Isomerases*: catalyse isomerization changes within a single molecule
- EC 6 *Ligases*: join two molecules with covalent bonds.

According to the naming conventions, enzymes are generally classified into six main family classes and many sub-family classes. Some web-servers, e.g., EzyPred and bioinformatics tools have been developed to predict which main family class and sub-family class an enzyme molecule belongs to according to its sequence information alone via the pseudo amino acid composition.

The Water Soluble Vitamins

A vitamin is an organic compound required as a nutrient in tiny amounts by an organism. In other words, an organic chemical compound (or related set of compounds) is called a vitamin when it cannot be synthesized in sufficient quantities by an organism, and must be obtained from the diet. Thus, the term is conditional both on the circumstances and the particular organism.

For example, ascorbic acid (vitamin C) is a vitamin for humans, but not for most other animals, and biotin and vitamin D are required in the human diet only in certain circumstances. By convention, the term *vitamin* does not include other essential nutrients such as dietary minerals, essential fatty acids or essential amino acids (which are needed in larger amounts than vitamins), nor does it encompass the large number of other nutrients that promote health, but are otherwise required less often. Thirteen vitamins are presently universally recognized.

Vitamins are classified by their biological and chemical activity, not their structure. Thus, each "vitamin" refers to a number of *vitamer* compounds that all show the biological activity associated with a particular vitamin. Such a set of chemicals are grouped under an alphabetized vitamin "generic descriptor" title, such as "vitamin A", which includes the compounds retinal, retinol, and four known carotenoids. Vitamers by definition are convertible to the active form of the vitamin in the body, and are sometimes inter-convertible to one another, as well.

Vitamins have diverse biochemical functions. Some have hormone-like functions as regulators of mineral metabolism (e.g. vitamin D), or regulators of cell and tissue growth and differentiation (e.g. some forms of vitamin A). Others function as antioxidants (e.g. vitamin E and sometimes vitamin C). The largest number of vitamins (e.g. B complex vitamins) function as precursors for enzyme cofactors, that help enzymes in their work as catalysts in metabolism.

In this role, vitamins may be tightly bound to enzymes as part of prosthetic groups: for example, biotin is part of enzymes involved in making fatty acids. Alternately, vitamins may also be less tightly bound to enzyme catalysts as coenzymes, detachable molecules which function to carry chemical groups or electrons between molecules. For example, folic acid carries various forms of carbon group – methyl, formyl and methylene – in the cell. Although these roles in assisting enzyme-substrate reactions are vitamins' best-known function, the other vitamin functions are equally important.

Until the mid-1930s, when the first commercial yeast-extract and semi-synthetic vitamin C supplement tablets were sold, vitamins were obtained solely through food intake, and changes in diet (which, for example, could occur during a particular growing season) can alter the types and amounts of vitamins ingested. Vitamins have been produced as commodity chemicals and made widely available as inexpensive semisynthetic and synthetic-source multivitamin dietary supplements, since the middle of the 20th century.

The term vitamin was historically derived from "vitamine," a combination word from *vita* and *amine,* meaning amine of life, because it was suggested in 1912 that the organic micronutrient food factors which prevented beriberi and perhaps other similar dietary-deficiency diseases, might be chemical amines. This proved incorrect for the micronutrient class, and the word was shortened to vitamin.

Vitamins are classified as either water-soluble or fat soluble. In humans there are 13 vitamins: 4 fat-soluble (A, D, E and K) and 9 water-soluble (8 B vitamins and vitamin C). Water-soluble vitamins dissolve easily in water, and in general, are readily excreted from the body, to the degree that urinary output is a strong predictor of vitamin consumption. Because they are not readily stored, consistent daily intake is important. Many types of water-soluble vitamins are synthesized by bacteria. Fat-soluble vitamins are absorbed through the intestinal tract with the help of lipids (fats). Because they are more likely to accumulate in the body, they are more likely to lead to hypervitaminosis than are water-soluble vitamins. Fat-soluble vitamin regulation is of particular significance in cystic fibrosis.

In Nutrition and Diseases

Vitamins are essential for the normal growth and development of a multicellular organism. Using the genetic blueprint inherited from its parents, a fetus begins to develop, at the moment of conception, from the nutrients it absorbs. It requires certain vitamins and minerals to be present at certain times. These nutrients facilitate the chemical reactions that produce among other things, skin, bone, and muscle. If there is serious deficiency in one or more of these nutrients, a child may develop a deficiency disease. Even minor deficiencies may cause permanent damage. For the most part, vitamins are obtained with food, but a few are obtained by other means. For example, microorganisms in the intestine—commonly known as "gut flora"—produce vitamin K and biotin, while one form of vitamin D is synthesized in the skin with the help of the natural ultraviolet wavelength of sunlight. Humans can produce some vitamins from precursors they consume. Examples include vitamin A, produced from beta carotene, and niacin, from the amino acid tryptophan. Once growth and development are completed, vitamins remain essential nutrients for the healthy maintenance of the cells, tissues, and organs that make up a multicellular organism; they also enable a multicellular life form to efficiently use chemical energy provided by food it eats, and to help process the proteins, carbohydrates, and fats required for respiration.

Deficiencies

It was suggested that when about 500 million years ago plants and animals began to transfer from the sea to rivers and land, environmental deficiency of marine mineral antioxidants, was a challenge to the evolution of terrestrial life. Terrestrial plants slowly optimized the production of "new" endogenous antioxidants such as ascorbic acid (Vitamin C), polyphenols, flavonoids, tocopherols, etc. Since this age dietary vitamin deficiencies appeared in terrestrial animals. Humans must consume vitamins periodically but with differing schedules, to avoid deficiency. Human bodily stores for different vitamins vary widely; vitamins A, D, and B_{12} are stored in significant amounts in the human body, mainly in the liver, and an adult human's diet may be deficient in vitamins A for many s and B_{12} in some cases for years, before developing a deficiency condition. However, vitamin B_3 (niacin and niacinamide) is not stored in the human body in significant amounts, so stores may only last a couple of weeks. For vitamin C, the first symptoms of scurvy in experimental studies of complete vitamin C deprivation in humans have varied widely, from a month to more than six months, depending on previous dietary history which determined body stores.

Deficiencies of vitamins are classified as either primary or secondary. A primary deficiency occurs when an organism does not get enough of the vitamin in its food. A secondary deficiency may be due to an underlying disorder that prevents or limits the absorption or use of the vitamin, due to a "lifestyle factor", such as smoking, excessive alcohol consumption, or the use of medications that interfere with the absorption or use of the vitamin. People who eat a varied diet are unlikely to develop a severe primary vitamin deficiency. In contrast, restrictive diets have the potential to cause prolonged vitamin deficits, which may result in often painful and potentially deadly diseases. Well-known human vitamin deficiencies involve thiamine (beriberi), niacin (pellagra), vitamin C (scurvy) and vitamin D (rickets). In much of the developed world, such deficiencies are rare; this is due to (1) an adequate supply of food; and (2) the addition of vitamins and minerals to common foods, often called fortification. In addition to these classical vitamin deficiency diseases, some evidence has also suggested links between vitamin deficiency and a number of different disorders.

Side Effects and Overdose

In large doses, some vitamins have documented side effects that tend to be more severe with a larger dosage. The likelihood of consuming too much of any vitamin from food is remote, but overdosing (vitamin poisoning) from vitamin supplementation does occur. At high enough dosages some vitamins cause side effects such as nausea, diarrhea, and vomiting. When side effects emerge, recovery is often accomplished by reducing the dosage. The doses of vitamins different individual can tolerate varies widely, and appear to be related to age and state of health. In 2008, overdose exposure to all formulations of vitamins and multivitamin-mineral formulations was reported by 68,911 individuals to the American Association of Poison Control Centres (nearly 80% of these exposures were in children under the age of 6), leading to 8 "major" life-threatening outcomes and 0 deaths.

Supplements

Dietary supplements, often containing vitamins, are used to ensure that adequate amounts of nutrients are obtained on a daily basis, if optimal amounts of the nutrients cannot be obtained through a varied diet. Scientific evidence supporting the benefits of some vitamin supplements is well established for certain health conditions, but others need further study. In some cases, vitamin supplements may have unwanted effects, especially if taken before surgery, with other dietary supplements or medicines, or if the person taking them has certain health conditions. Dietary supplements may also contain levels of vitamins many times higher, and in different forms, than one may ingest through food. There have been mixed studies on the importance and safety of dietary supplementation.

A meta-analysis published in 2006 suggested that Vitamin A and E supplements not only provide no tangible health benefits for generally healthy individuals, but may actually increase mortality, although two large studies included in the analysis involved smokers, for which it was already known that beta-carotene supplements can be harmful. Another study published in May 2009 found that antioxidants such as vitamins C and E may actually curb some benefits of exercise. While others findings suggest that evidence of Vitamin E toxicity is limited to specific form taken in excess.

The water-soluble vitamins, excluding vitamin C, popularly are termed the B-complex vitamins. There are eight of them, namely; B1 (thiamine), B2 (riboflavin), B6 (pyridoxine), niacin (nicotinic acid), B12, folic acid, pantothenic acid, and biotin. The water-soluble vitamins, inactive in their so-called free states, must be activated to their coenzyme forms. B-complex vitamins and vitamin C are water-soluble vitamins that are not stored in the body and must be replaced each day, preferably through a high-quality liquid multivitamin.

The water-soluble vitamins are absorbed in our intestine, pass directly to the blood, and are carried to the tissues in which they will be utilized. Vitamin B12 requires a substance known as intrinsic factor for absorption". Water-soluble vitamins usually are excreted in the urine on a daily basis. Thiamine (B1), riboflavin (B2), pyridoxine (B6), ascorbic acid (C), pantothenic acid, and biotin appear in urine as free vitamins Tissue storage capacity of water soluble vitamins is limited and, as the tissues become saturated, the rate of excretion increases sharply. This keeps us from overdosing but this is also why we need to take these vitamins daily. Unlike the other water-soluble vitamins, however, vitamin B12 is excreted solely in the feces. Some folic acid and biotin is also normally excreted in this way. Although fecal excretion of water-soluble vitamins (other than vitamin B12, folic acid, and biotin) occurs, their source probably is the intestinal bacteria, which synthesize the vitamins, rather than vitamins that we have eaten and used.

The effects of the water-soluble vitamins are obvious in many parts of the body. They act as coenzymes to help the body obtain energy from food. They also are important for normal appetite, good vision, healthy skin, healthy nervous system and red blood cell formation. Vitamin B deficiency leads to beriberi, pellagra and pernicious anemia. Alcoholics are especially prone to thiamin deficiency because they hardly eat good food or any at all. You can also become deficient in vitamin B if you eat polished rice, that is, removing its outer layer, and end up with thiamin deficiency, or beriberi. Like the other vitamins, the body needs vitamin C to keep it in good working condition. Also called ascorbic acid, vitamin C helps hold body cells together, aids in wound healing, assists in bone and tooth formation, and strengthens the blood vessel walls.

Vitamin C is also crucial to the functioning of our immune system, and it helps improve the absorption and utilization of iron. Vitamin C also helps prevent scurvy. Our bodies cannot make vitamin C and our capacity to store vitamin C is limited. We must, therefore, take in some daily. You are likely to need even more vitamin C if you are under stress, use oral contraceptives, have a healing wound, if you are pregnant or a child, have fever or infections, and if you smoke cigarettes. Megadoses of vitamin C can be help prevent or possibly even cure a case of the common cold.

Vitamin C also serves as a powerful antioxidant. It works synergistically with vitamin E as a free-radical scavenger. Studies suggest that vitamin C may reduce the risk of certain cancers, heart disease and cataracts. Recent studies also suggest that the combination of vitamins C and E in high doses can help reduce the risk of developing Alzheimers disease. Vitamins C and E can both be found in a high-quality liquid multivitamin. Remember that water-soluble vitamins, the B-complex group and vitamin C, dissolve in water, are not stored and they are eliminated in urine. We need a continuous supply of them in our diets to ensure proper health and nutrition.

Chapter

6

Dietary Antioxidants

An antioxidant is a molecule capable of inhibiting the oxidation of other molecules. Oxidation is a chemical reaction that transfers electrons from a substance to an oxidizing agent. Oxidation reactions can produce free radicals. In turn, these radicals can start chain reactions that damage cells. Antioxidants terminate these chain reactions by removing free radical intermediates, and inhibit other oxidation reactions. They do this by being oxidized themselves, so antioxidants are often reducing agents such as thiols, ascorbic acid or polyphenols. Although oxidation reactions are crucial for life, they can also be damaging; hence, plants and animals maintain complex systems of multiple types of antioxidants, such as glutathione, vitamin C, and vitamin E as well as enzymes such as catalase, superoxide dismutase and various peroxidases.

Low levels of antioxidants, or inhibition of the antioxidant enzymes, cause oxidative stress and may damage or kill cells. As oxidative stress might be an important part of many human diseases, the use of antioxidants in pharmacology is intensively studied, particularly as treatments for stroke and neurodegenerative diseases. However, it is unknown whether oxidative stress is the cause or the consequence of disease. Antioxidants are widely used as ingredients in dietary supplements in the hope of maintaining health and preventing diseases such as cancer, coronary heart disease and even altitude sickness. Although initial studies suggested that antioxidant supplements might promote health, later large clinical trials did not detect any benefit and suggested instead that excess supplementation may be harmful. In addition to these uses of natural antioxidants in medicine, these compounds have many industrial uses, such as preservatives in food and cosmetics and preventing the degradation of rubber and gasoline.

As part of their adaptation from marine life, terrestrial plants began producing non-marine antioxidants such as ascorbic acid (Vitamin C), polyphenols, flavonoids and tocopherols. Further development of angiosperm plants between 50 and 200 million years ago, particularly during the Jurassic period, produced many antioxidant pigments evolved during the late Jurassic period as chemical defences against reactive oxygen species produced during photosynthesis. The term antioxidant originally was used to refer specifically to a chemical that prevented the consumption of oxygen. In the late 19th century and early 20th century, extensive study was devoted to the uses of antioxidants in important industrial processes, such as the prevention of metal corrosion, the vulcanization of rubber, and the polymerization of fuels in the fouling of internal combustion engines.

Early research on the role of antioxidants in biology focused on their use in preventing the oxidation of unsaturated fats, which is the cause of rancidity. Antioxidant activity could be measured simply by placing the fat in a closed container with oxygen and measuring the rate of oxygen consumption. However, it was the identification of vitamins A, C, and E as antioxidants that revolutionized the field and led to the realization of the importance of antioxidants in the biochemistry of living organisms. The possible mechanisms of action of antioxidants were first explored when it was recognized that a substance with anti-oxidative activity is likely to be one that is itself readily oxidized. Research into how vitamin E prevents the process of lipid peroxidation led to the identification of antioxidants as reducing agents that prevent oxidative reactions, often by scavenging reactive oxygen species before they can damage cells.

The Oxidative Challenge in Biology

A paradox in metabolism is that while the vast majority of complex life on Earth requires oxygen for its existence, oxygen is a highly reactive molecule that damages living organisms by producing reactive oxygen species. Consequently, organisms contain a complex network of antioxidant metabolites and enzymes that work together to prevent oxidative damage to cellular components such as DNA, proteins and lipids. In general, antioxidant systems either prevent these reactive species from being formed, or remove them before they can damage vital components of the cell. However, since reactive oxygen species do have useful functions in cells, such as redox signalling, the function of antioxidant systems is not to remove oxidants entirely, but instead to keep them at an optimum level. The reactive oxygen species produced in cells include hydrogen peroxide (H_2O_2), hypochlorous acid (HOCl), and free radicals such as the hydroxyl radical (·OH) and the superoxide anion (O_2^-). The hydroxyl radical is particularly unstable and will react rapidly and non-specifically with most biological molecules.

This species is produced from hydrogen peroxide in metal-catalysed redox reactions such as the Fenton reaction. These oxidants can damage cells by starting chemical chain reactions such as lipid peroxidation, or by oxidizing DNA or proteins. Damage to DNA can cause mutations and possibly cancer, if not reversed by DNA repair mechanisms, while damage to proteins causes enzyme inhibition, denaturation and protein degradation. The use of oxygen as part of the process for generating metabolic energy produces reactive oxygen species. In this process, the superoxide anion is produced as a by-product of several steps in the electron transport chain. Particularly important is the reduction of coenzyme Q in complex III, since a highly reactive free radical is formed as an intermediate (Q ·−).

This unstable intermediate can lead to electron "leakage", when electrons jump directly to oxygen and form the superoxide anion, instead of moving through the normal series of well-controlled reactions of the electron transport chain. Peroxide is also produced from the oxidation of reduced flavoproteins, such as complex I. However, although these enzymes can produce oxidants, the relative importance of the electron transfer chain to other processes that generate peroxide is unclear. In plants, algae, and cyanobacteria, reactive oxygen species are also produced during photosynthesis, particularly under conditions of high light intensity. This effect is partly offset by the involvement of carotenoids in photoinhibition, which involves these antioxidants reacting with over-reduced forms of the photosynthetic reaction centres to prevent the production of reactive oxygen species.

Metabolites

Antioxidants are classified into two broad divisions, depending on whether they are soluble in water (hydrophilic) or in lipids (hydrophobic). In general, water-soluble antioxidants react with oxidants in the cell cytosol and the blood plasma, while lipid-soluble antioxidants protect cell membranes from lipid peroxidation. These compounds may be synthesized in the body or obtained from the diet. The different antioxidants are present at a wide range of concentrations in body fluids and tissues, with some such as glutathione or ubiquinone mostly present within cells, while others such as uric acid are more evenly distributed. Some antioxidants are only found in a few organisms and these compounds can be important in pathogens and can be virulence factors.

The relative importance and interactions between these different antioxidants is a very complex question, with the various metabolites and enzyme systems having synergistic and interdependent effects on one another. The action of one antioxidant may therefore depend on the proper function of other members of the antioxidant system. The amount of protection provided by any one antioxidant will also depend on its concentration, its reactivity towards the particular reactive oxygen species being considered, and the status of the antioxidants with which it interacts.

Some compounds contribute to antioxidant defence by chelating transition metals and preventing them from catalysing the production of free radicals in the cell. Particularly important is the ability to sequester iron, which is the function of iron-binding proteins such as transferrin and ferritin. Selenium and zinc are commonly referred to as *antioxidant nutrients*, but these chemical elements have no antioxidant action themselves and are instead required for the activity of some antioxidant enzymes.

Uric Acid

The antioxidant in highest concentration in human blood is uric acid, which provides about half of the total antioxidant capacity of human serum. Uric acid is an oxypurine produced from xanthine by the enzyme xanthine oxidase, and is a waste product of purine metabolism in primates, birds, and reptiles. An overabundance of this chemical in the body causes gout. The effects of uric acid in conditions such as stroke and heart attacks are still not well understood, with some studies linking higher levels of uric acid with increased mortality. This apparent effect might either be due to uric acid being activated as a defence mechanism against oxidative stress, or uric acid acting as a pro-oxidant and contributing to the damage caused in these diseases. Uric acid is released from tissues that are short of oxygen and elevated uric acid levels may be an important part of acclimatisation to high altitude.

Ascorbic Acid

Ascorbic acid or “vitamin C” is a monosaccharide oxidation-reduction (redox) catalyst found in both animals and plants. As one of the enzymes needed to make ascorbic acid has been lost by mutation during primate evolution, humans must obtain it from the diet; it is therefore a vitamin. Most other animals are able to produce this compound in their bodies and do not require it in their diets.

Ascorbic acid is required for the conversion of the procollagen to collagen by oxidizing proline residues to hydroxyproline. In other cells, it is maintained in its reduced form by reaction with glutathione, which can be catalysed by protein disulfide isomerase and glutaredoxins. Ascorbic acid is redox catalyst which can reduce, and thereby neutralize, reactive oxygen species such as hydrogen peroxide. In addition to its direct antioxidant effects, ascorbic acid is also a substrate for the redox enzyme ascorbate peroxidase, a function that is particularly important in stress resistance in plants. Ascorbic acid is present at high levels in all parts of plants and can reach concentrations of 20 millimolar in chloroplasts.

Glutathione

Glutathione is a cysteine-containing peptide found in most forms of aerobic life. It is not required in the diet and is instead synthesized in cells from its constituent amino acids. Glutathione has antioxidant properties since the thiol group in its cysteine moiety is a reducing agent and can be reversibly oxidized and reduced.

In cells, glutathione is maintained in the reduced form by the enzyme glutathione reductase and in turn reduces other metabolites and enzyme systems, such as ascorbate in the glutathione-ascorbate cycle, glutathione peroxidases and glutaredoxins, as well as reacting directly with oxidants. Due to its high concentration and its central role in maintaining the cell’s redox state, glutathione is one of the most important cellular antioxidants. In some organisms glutathione is replaced by other thiols, such as by mycothiol in the Actinomycetes, or by trypanothione in the Kinetoplastids.

Melatonin

Melatonin is a powerful antioxidant and, unlike conventional antioxidants such as vitamins C and E and glutathione, it is both produced in the human body and is acquired in the diet (fruits, vegetables, cereals and herbs etc., contain melatonin). Melatonin easily crosses cell membranes and the blood-brain barrier Unlike other antioxidants, melatonin does not undergo redox cycling, which is the ability of a molecule to undergo repeated reduction and oxidation.

Redox cycling may allow other antioxidants (such as vitamin C) to act as pro-oxidants and promote free radical formation. Melatonin, once oxidized, cannot be reduced to its former state because it forms several stable end-products upon reacting with free radicals. Therefore, it has been referred to as a terminal (or suicidal) antioxidant.

Tocopherols and Tocotrienols (Vitamin E)

Vitamin E is the collective name for a set of eight related tocopherols and tocotrienols, which are fat-soluble vitamins with antioxidant properties. Of these, α-tocopherol has been most studied as it has the highest bioavailability, with the body preferentially absorbing and metabolising this form.

It has been claimed that the α-tocopherol form is the most important lipid-soluble antioxidant, and that it protects membranes from oxidation by reacting with lipid radicals produced in the lipid peroxidation chain reaction. This removes the free radical intermediates and prevents the propagation reaction from continuing. This reaction produces oxidised α-tocopheroxyl radicals that can be recycled back to the active reduced form through reduction by other antioxidants, such as ascorbate, retinol or ubiquinol.

This is in line with findings showing that α-tocopherol, but not water-soluble antioxidants, efficiently protects glutathione peroxidase 4 (GPX4)-deficient cells from cell death. GPx4 is the only known enzyme that efficiently reduces lipid-hydroperoxides within biological membranes.

However, the roles and importance of the various forms of vitamin E are presently unclear, and it has even been suggested that the most important function of α-tocopherol is as a signalling molecule, with this molecule having no significant role in antioxidant metabolism. The functions of the other forms of vitamin E are even less well-understood, although γ-tocopherol is a nucleophile that may react with electrophilic mutagens, and tocotrienols may be important in protecting neurons from damage.

Vitamin E

Vitamin E is a generic term for tocopherols and tocotrienols. Vitamin E is a family of α-, β-, γ-, and δ- (respectively: alpha, beta, gamma, and delta) tocopherols and corresponding four tocotrienols. Vitamin

E is a fat-soluble antioxidant that stops the production of reactive oxygen species formed when fat undergoes oxidation. Of these, α-tocopherol (also written as alpha-tocopherol) has been most studied as it has the highest bioavailability.

α-Tocopherol

It has been claimed that α-tocopherol is the most important lipid-soluble antioxidant, and that it protects cell membranes from oxidation by reacting with lipid radicals produced in the lipid peroxidation chain reaction. This would remove the free radical intermediates and prevent the oxidation reaction from continuing. The oxidised α-tocopheroxyl radicals produced in this process may be recycled back to the active reduced form through reduction by other antioxidants, such as ascorbate, retinol or ubiquinol.

However, the importance of the antioxidant properties of this molecule at the concentrations present in the body are not clear and it is possible that the reason why vitamin E is required in the diet is unrelated to its ability to act as an antioxidant. Other forms of vitamin E have their own unique properties. For example, γ-tocopherol (also written as gamma-tocopherol) is a nucleophile that can react with electrophilic mutagens. However, the roles and importance of all of the various forms of vitamin E are presently unclear, and it has even been suggested that the most important function of vitamin E is as a signalling molecule, and that it has no significant role in antioxidant metabolism. So far, most studies about vitamin E have supplemented using only alpha-tocopherol, but doing so leads to reduced serum gamma- and delta-tocopherol concentrations. Moreover, a 2007 clinical study involving alpha-tocopherol concluded that supplementation did not reduce the risk of major cardiovascular events in middle aged and older men.

Tocotrienols

Compared with tocopherols, tocotrienols are sparsely studied. Less than 1% of PubMed papers on vitamin E relate to tocotrienols. Current research direction is starting to give more prominence to the tocotrienols, the lesser known but more potent antioxidants in the vitamin E family. Some studies have suggested that tocotrienols have specialized roles in protecting neurons from damage and cholesterol reduction by inhibiting the activity of HMG-CoA reductase[16-1];δ-tocotrienol blocks processing of sterol regulatory element binding proteins (SREBPs)[16-1]. Oral consumption of tocotrienols is also thought to protect against stroke-associated brain damage in vivo.

Until further research has been carried out on the other forms of vitamin E, conclusions relating to the other forms of vitamin E, based on trials studying only the efficacy of α-tocopherol, may be premature.

Recommended Daily Intake

The Food and Nutrition Board at the Institute of Medicine report the following dietary reference intakes for vitamin E:

Infants:

- 0 to 6 months: 4 mg/day
- 7 to 12 months: 5 mg/day.

Children:

- 1 to 3 years: 6 mg/day
- 4 to 8 years: 7 mg/day
- 9 to 13 years: 11 mg/day.

Adolescents and Adults:

- 14 and older: 15 mg/day.

Dietary Sources and Supplements

The following foods are rich in vitamin E:

- fortified cereals
- seeds and seed oils, like sunflower
- nuts and nut oils, like almonds and hazelnuts
- green leafy vegetables, like spinach, turnip, beet, collard, and dandelion greens
- tomato products
- pumpkin
- sweet potato
- blue crab
- rockfish
- mangoes
- asparagus
- broccoli
- papayas.

Health Effects

The consensus in the medical community is that there is no good evidence to support health benefits from vitamin E supplementation, yet there is strong evidence that taking more than 400 IU of vitamin E per day for extended periods increases the risk of death.

Pro-oxidant

Pro-oxidants are chemicals that induce oxidative stress, either through creating reactive oxygen species or inhibiting antioxidant systems. The oxidative stress produced by these chemicals can damage cells and tissues, for example an overdose of the analgesic paracetamol (acetaminophen) can cause fatal damage to the liver, partly through its production of reactive oxygen species. Some substances can act as either antioxidants, or pro-oxidants, depending on the specific set of conditions. Some of the conditions that are important include the concentration of the chemical and if oxygen or transition metals are present. While thermodynamically very favoured, reduction of molecular oxygen or peroxide to superoxide or hydroxyl radical is fortunately spin forbidden. This greatly reduces the rates of these reactions, thus allowing aerobic life to exist. As a result, the reduction of oxygen typically involves either the initial formation of singlet oxygen, or spin-orbit coupling through a reduction of a transition-series metal such as manganese, iron, or copper. This reduced metal then transfers the single electron to molecular oxygen or peroxide.

Metals

Transition metals can act as pro-oxidants. E.g., chronic manganism is a classic “pro-oxidant” disease. Another disease associated with the chronic presence of a pro-oxidant transition-series metal is hemochromatosis, associated with elevated iron levels. Similarly, Wilson’s disease is associated with elevated tissue levels of copper. Such syndromes tend to be associated with a common symptomology. This typically-includes various combinations of psychosis, dyskinesia (including Parkinsonian-like symptomology), pigmentary abnormalities, fibrosis, deafness, diabetes, and arthritis. Thus, all are occasional symptoms of (e.g) hemochromatosis, another name for which is “bronze diabetes”. The pro-oxidant herbicide paraquat, Wilson’s disease, and striatal iron have similarly been linked to human Parkinsonism. Paraquat also produces parkinsonian-like symptoms in rodents.

Fibrosis

Fibrosis or scar formation is another pro-oxidant-related symptom. E.g., interocular copper or vitreous chalcosis is associated with severe vitreous fibrosis, as is interocular iron. Liver cirrhosis is also a major symptom of Wilson’s disease. The pulmonary fibrosis produced by paraquat and the antitumor agent bleomycin is also thought to be induced by the pro-oxidant properties of these agents. It may be that oxidative stress produced by such agents mimics a normal physiological signal for fibroblast conversion to myofibroblasts.

Pro-oxidant Vitamins

Vitamins that are reducing agents can be pro-oxidants. Vitamin C has antioxidant activity when it reduces oxidizing substances such as hydrogen peroxide, however, it can also reduce metal ions which leads to the generation of free radicals through the fenton reaction.

$$2\ Fe^{2+} + 2\ H_2O_2 \rightarrow 2\ Fe^{3+} + 2\ OH\cdot + 2\ OH^-$$

$$2\ Fe^{3+} + \text{Ascorbate} \rightarrow 2\ Fe^{2+} + \text{Dehydroascorbate.}$$

The metal ion in this reaction can be reduced, oxidized, and then re-reduced, in a process called redox cycling that can generate reactive oxygen species.

The relative importance of the antioxidant and pro-oxidant activities of antioxidant vitamins are an area of current research, but vitamin C, for example, appears to have a mostly antioxidant action in the body. However, less data is available for other dietary antioxidants, such as polyphenol antioxidants, zinc, and vitamin E.

Uric Acid

The pro-oxidant properties of reductants can also have clinical consequences. For example, in humans, uric acid accounts for roughly half the antioxidant ability of plasma. In fact, uric acid may have substituted for ascorbate in human evolution. However, like ascorbate, uric acid can also mediate the production of active oxygen species and thus act as a prooxidant.

This was first proposed to play a role in the etiology of the Lesch-Nyhan Syndrome (associated with choreoathetoid dyskinesia) and in hyperuricemic syndrome in dalmatian dogs. The latter responds to treatment with the antioxidant drug orgotein, a pharmaceutical form of superoxide dismutase. Such animals are also typically "bronzed".

High uric acid levels are also encountered in atherosclerosis, metabolic syndrome, and in stroke. The issue is whether hyperuricemia is a protective response to oxidative stress in such diseases or whether it is a primary cause.

Thus, some researchers think urate-induced oxidative stress is causative in stroke, while others suggest the exact opposite, that urate is neuroprotective by means of its antioxidant properties. Similarly, evidence relates metabolic syndrome to the pro-oxidant properties of urate secondary to fructose-induced hyperuricemia.

Hyperhomocysteinemia

In humans, elevated homocysteine levels in hyperhomocysteinemia are associated with an increased incidence of atherosclerosis and may play a role in Alzheimers. Homocysteine is a powerful reducing agent and like most such agents can induce oxidative stress through reducing molecular oxygen to its radical forms. The spin-forbidden nature of this reaction normally requires this be mediated through reduction of some heavy atom, which then transfers the electron to oxygen. Repeated working of this process is known as redox cycling. This has been proposed to play a role in the etiology of such diseases.

Anticancer Drugs

Several important anticancer agents both bind to DNA and generate reactive oxygen species. These include adriamycin and other anthracyclines, bleomycin, and cisplatin. These agents may show specific toxicity towards cancer cells because of the low level of antioxidant defenses found in tumours.

Recent research demonstrates that redox dysregulation originating from metabolic alterations and dependence on mitogenic and survival signalling through reactive oxygen species represents a specific vulnerability of malignant cells that can be selectively targeted by pro-oxidant non-genotoxic redox chemotherapeutics.

Pro-oxidant Activities

Antioxidants that are reducing agents can also act as pro-oxidants. For example, vitamin C has antioxidant activity when it reduces oxidizing substances such as hydrogen peroxide, however, it will also reduce metal ions that generate free radicals through the Fenton reaction.

$$2\ Fe^{3+} + \text{Ascorbate} \rightarrow 2\ Fe^{2+} + \text{Dehydroascorbate}$$

$$2\ Fe^{2+} + 2\ H_2O_2 \rightarrow 2\ Fe^{3+} + 2\ OH\cdot + 2\ OH^-$$

The relative importance of the antioxidant and pro-oxidant activities of antioxidants are an area of current research, but vitamin C, for example, appears to have a mostly antioxidant action in the body. However, less data is available for other dietary antioxidants, such as vitamin E, or the polyphenols.

Potential of Antioxidant Supplements to Damage Health

There is evidence antioxidant supplements may be disease-promoting and increase mortality in humans. It was previously proposed on a hypothetical basis that free radicals may induce an endogenous response culminating in more effective adaptations which protect against exogenous radicals (and possibly other toxic compounds). Recent experimental evidence strongly suggests that this is indeed the case, and that such induction of endogenous free radical production extends life span of a model organism. Most importantly, this induction of life span is prevented by antioxidants, providing direct evidence that toxic radicals may mitohormetically exert life extending and health promoting effects..

Enzyme Systems

As with the chemical antioxidants, cells are protected against oxidative stress by an interacting network of antioxidant enzymes. Here, the superoxide released by processes such as oxidative phosphorylation is first converted to hydrogen peroxide and then further reduced to give water. This detoxification pathway is the result of multiple enzymes, with superoxide dismutases catalysing the first step and then catalases and various peroxidases removing hydrogen peroxide. As with antioxidant metabolites, the contributions of these enzymes to antioxidant defenses can be hard to separate from one another, but the generation of transgenic mice lacking just one antioxidant enzyme can be informative.

Superoxide Dismutase, Catalase and Peroxiredoxins

Superoxide dismutases (SODs) are a class of closely related enzymes that catalyse the breakdown of the superoxide anion into oxygen and hydrogen peroxide. SOD enzymes are present in almost all aerobic cells and in extracellular fluids. Superoxide dismutase enzymes contain metal ion cofactors that, depending on the isozyme, can be copper, zinc, manganese or iron.

In humans, the copper/zinc SOD is present in the cytosol, while manganese SOD is present in the mitochondrion. There also exists a third form of SOD in extracellular fluids, which contains copper and zinc in its active sites. The mitochondrial isozyme seems to be the most biologically important of these three, since mice lacking this enzyme die soon after birth. In contrast, the mice lacking copper/zinc SOD (Sod1) are viable but have numerous pathologies and a reduced lifespan, while mice without the extracellular SOD have minimal defects (sensitive to hyperoxia). In plants, SOD isozymes are present in the cytosol and mitochondria, with an iron SOD found in chloroplasts that is absent from vertebrates and yeast.

Catalases are enzymes that catalyse the conversion of hydrogen peroxide to water and oxygen, using either an iron or manganese cofactor. This protein is localized to peroxisomes in most eukaryotic cells. Catalase is an unusual enzyme since, although hydrogen peroxide is its only substrate, it follows a ping-pong mechanism.

Here, its cofactor is oxidised by one molecule of hydrogen peroxide and then regenerated by transferring the bound oxygen to a second molecule of substrate. Despite its apparent importance in hydrogen peroxide removal, humans with genetic deficiency of catalase — "acatalasemia" — or mice genetically engineered to lack catalase completely, suffer few ill effects.

Peroxiredoxins are peroxidases that catalyse the reduction of hydrogen peroxide, organic hydroperoxides, as well as peroxynitrite. They are divided into three classes: typical 2-cysteine peroxiredoxins; atypical 2-cysteine peroxiredoxins; and 1-cysteine peroxiredoxins. These enzymes share the same basic catalytic mechanism, in which a redox-active cysteine (the peroxidatic cysteine) in the active site is oxidized to a sulfenic acid by the peroxide substrate.

Over-oxidation of this cysteine residue in peroxiredoxins inactivates these enzymes, but this can be reversed by the action of sulfiredoxin. Peroxiredoxins seem to be important in antioxidant metabolism, as mice lacking peroxiredoxin 1 or 2 have shortened lifespan and suffer from hemolytic anaemia, while plants use peroxiredoxins to remove hydrogen peroxide generated in chloroplasts.

Thioredoxin and Glutathione Systems

The thioredoxin system contains the 12-kDa protein thioredoxin and its companion thioredoxin reductase. Proteins related to thioredoxin are present in all sequenced organisms with plants, such as *Arabidopsis thaliana,* having a particularly great diversity of isoforms.

The active site of thioredoxin consists of two neighbouring cysteines, as part of a highly conserved CXXC motif, that can cycle between an active dithiol form (reduced) and an oxidized disulfide form. In its active state, thioredoxin acts as an efficient reducing agent, scavenging reactive oxygen species and maintaining other proteins in their reduced state. After being oxidized, the active thioredoxin is regenerated by the action of thioredoxin reductase, using NADPH as an electron donor. The glutathione system includes glutathione, glutathione reductase, glutathione peroxidases and glutathione *S*-transferases.

This system is found in animals, plants and microorganisms. Glutathione peroxidase is an enzyme containing four selenium-cofactors that catalyzes the breakdown of hydrogen peroxide and organic hydroperoxides. There are at least four different glutathione peroxidase isozymes in animals.

Glutathione peroxidase 1 is the most abundant and is a very efficient scavenger of hydrogen peroxide, while glutathione peroxidase 4 is most active with lipid hydroperoxides. Surprisingly, glutathione peroxidase 1 is dispensable, as mice lacking this enzyme have normal lifespans, but they are hypersensitive to induced oxidative stress. In addition, the glutathione *S*-transferases show high activity with lipid peroxides. These enzymes are at particularly high levels in the liver and also serve in detoxification metabolism.

Oxidative Stress in Disease

Oxidative stress is thought to contribute to the development of a wide range of diseases including Alzheimer's disease, Parkinson's disease, the pathologies caused by diabetes, rheumatoid arthritis, and neurodegeneration in motor neuron diseases. In many of these cases, it is unclear if oxidants trigger the disease, or if they are produced as a secondary consequence of the disease and from general tissue damage; One case in which this link is particularly well-understood is the role of oxidative stress in cardiovascular disease. Here, low density lipoprotein (LDL) oxidation appears to trigger the process of atherogenesis, which results in atherosclerosis, and finally cardiovascular disease.

A low calorie diet extends median and maximum lifespan in many animals. This effect may involve a reduction in oxidative stress. While there is some evidence to support the role of oxidative stress in aging in model organisms such as *Drosophila melanogaster* and *Caenorhabditis elegans*, the evidence in mammals is less clear.

Indeed, a 2009 review of experiments in mice concluded that almost all manipulations of antioxidant systems had no effect on aging. Diets high in fruit and vegetables, which are high in antioxidants, promote health and reduce the effects of aging, however antioxidant vitamin supplementation has no detectable effect on the aging process, so the effects of fruit and vegetables may be unrelated to their antioxidant contents. One reason for this might be the fact that consuming antioxidant molecules such as polyphenols and vitamin E will produce changes in other parts of metabolism, so it may be these other effects that are the real reason these compounds are important in human nutrition.

Health Effects

Disease Treatment

The brain is uniquely vulnerable to oxidative injury, due to its high metabolic rate and elevated levels of polyunsaturated lipids, the target of lipid peroxidation. Consequently, antioxidants are commonly used as medications to treat various forms of brain injury. Here, superoxide dismutase mimetics, sodium thiopental and propofol are used to treat reperfusion injury and traumatic brain injury, while the experimental drug NXY-059 and ebselen are being applied in the treatment of stroke. These compounds appear to prevent oxidative stress in neurons and prevent apoptosis and neurological damage. Antioxidants are also being investigated as possible treatments for neurodegenerative diseases such as Alzheimer's disease, Parkinson's disease, and amyotrophic lateral sclerosis, and as a way to prevent noise-induced hearing loss.

Disease Prevention

People who eat fruits and vegetables have a lower risk of heart disease and some neurological diseases, and there is evidence that some types of vegetables, and fruits in general, protect against some cancers. Since fruits and vegetables happen to be good sources of antioxidants, this suggested that antioxidants might prevent some types of diseases.

This idea has been tested in clinical trials and does not seem to be true, as antioxidant supplements have no clear effect on the risk of chronic diseases such as cancer and heart disease. This suggests that these health benefits come from other substances in fruits and vegetables (possibly flavonoids), or come from a complex mix of substances.

It is thought that oxidation of low density lipoprotein in the blood contributes to heart disease, and initial observational studies found that people taking Vitamin E supplements had a lower risk of developing heart disease. Consequently, at least seven large clinical trials were conducted to test the effects of antioxidant supplement with Vitamin E, in doses ranging from 50 to 600 mg per day. None of these trials found a statistically significant effect of Vitamin E on overall number of deaths or on deaths due to heart disease. Further studies have also been negative.

It is not clear if the doses used in these trials or in most dietary supplements are capable of producing any significant decrease in oxidative stress. Overall, despite the clear role of oxidative stress in cardiovascular disease, controlled studies using antioxidant vitamins have observed no reduction in either the risk of developing heart disease, or the rate of progression of existing disease.

While several trials have investigated supplements with high doses of antioxidants, the "*Supplémentation en Vitamines et Mineraux Antioxydants*" (SU.VI.MAX) study tested the effect of supplementation with doses comparable to those in a healthy diet. Over 12,500 French men and women took either low-dose antioxidants (120 mg of ascorbic acid, 30 mg of vitamin E, 6 mg of beta carotene, 100 µg of selenium, and 20 mg of zinc) or placebo pills for an average of 7.5 years. The investigators found there was no statistically significant effect of the antioxidants on overall survival, cancer, or heart disease. In a post-hoc analysis they found a 31% reduction in the risk of cancer in men, but not women.

Many nutraceutical and health food companies sell formulations of antioxidants as dietary supplements and these are widely used in industrialized countries. These supplements may include specific antioxidant chemicals, like the polyphenol, resveratrol (from grape seeds or knotweed roots), combinations of antioxidants, like the "ACES" products that contain beta carotene (provitamin A), vitamin C, vitamin E and Selenium, or herbs that contain antioxidants-such as green tea and jiaogulan.

Although some levels of antioxidant vitamins and minerals in the diet are required for good health, there is considerable doubt as to whether these antioxidant supplements are beneficial or harmful, and if they are actually beneficial, which antioxidant(s) are needed and in what amounts. Indeed, some authors argue that the hypothesis that antioxidants could prevent chronic diseases has now been disproved and that the idea was misguided from the beginning. Rather, dietary polyphenols may have non-antioxidant roles in minute concentrations that affect cell-to-cell signalling, receptor sensitivity, inflammatory enzyme activity or gene regulation.

For overall life expectancy, it has even been suggested that moderate levels of oxidative stress may increase lifespan in the worm *Caenorhabditis elegans*, by inducing a protective response to increased levels of reactive oxygen species. The suggestion that increased life expectancy comes from increased oxidative stress conflicts with results seen in the yeast *Saccharomyces cerevisiae*, and the situation in mammals is even less clear. Nevertheless, antioxidant supplements do not appear to increase life expectancy in humans.

Physical Exercise

During exercise, oxygen consumption can increase by a factor of more than 10. This leads to a large increase in the production of oxidants and results in damage that contributes to muscular fatigue during and after exercise. The inflammatory response that occurs after strenuous exercise is also associated with oxidative stress, especially in the 24 hours after an exercise session. The immune system response to the damage done by exercise peaks 2 to 7 days after exercise, which is the period during which most of the adaptation that leads to greater fitness occurs. During this process, free radicals are produced by neutrophils to remove damaged tissue. As a result, excessive antioxidant levels may inhibit recovery and adaptation mechanisms. Antioxidant supplements may also prevent any of the health gains that normally come from exercise, such as increased insulin sensitivity.

The evidence for benefits from antioxidant supplementation in vigorous exercise is mixed. There is strong evidence that one of the adaptations resulting from exercise is a strengthening of the body's antioxidant defenses, particularly the glutathione system, to regulate the increased oxidative stress. This effect may be to some extent protective against diseases which are associated with oxidative stress, which would provide a partial explanation for the lower incidence of major diseases and better health of those who undertake regular exercise.

However, no benefits for physical performance to athletes are seen with vitamin E supplementation. Indeed, despite its key role in preventing lipid membrane peroxidation, 6 weeks of vitamin E supplementation had no effect on muscle damage in ultramarathon runners. Although there appears to be no increased requirement for vitamin C in athletes, there is some evidence that vitamin C supplementation increased the amount of intense exercise that can be done and vitamin C supplementation before strenuous exercise may reduce the amount of muscle damage. However, other studies found no such effects, and some research suggests that supplementation with amounts as high as 1000 mg inhibits recovery.

Adverse Effects

Relatively strong reducing acids can have antinutrient effects by binding to dietary minerals such as iron and zinc in the gastrointestinal tract and preventing them from being absorbed. Notable examples are oxalic acid, tannins and phytic acid, which are high in plant-based diets. Calcium and iron deficiencies are not uncommon in diets in developing countries where less meat is eaten and there is high consumption of phytic acid from beans and unleavened whole grain bread. Nonpolar antioxidants such as eugenol—a major component of oil of cloves—have toxicity limits that can be exceeded with the misuse of undiluted essential oils.

Toxicity associated with high doses of water-soluble antioxidants such as ascorbic acid are less of a concern, as these compounds can be excreted rapidly in urine. More seriously, very high doses of some antioxidants may have harmful long-term effects. The beta-Carotene and Retinol Efficacy Trial (CARET) study of lung cancer patients found that smokers given supplements containing beta-carotene and vitamin A had increased rates of lung cancer. Subsequent studies confirmed these adverse effects.

These harmful effects may also be seen in non-smokers, as a recent meta-analysis including data from approximately 230,000 patients showed that β-carotene, vitamin A or vitamin E supplementation is associated with increased mortality but saw no significant effect from vitamin C. No health risk was seen when all the randomized controlled studies were examined together, but an increase in mortality was detected only when the high-quality and low-bias risk trials were examined separately.

However, as the majority of these low-bias trials dealt with either elderly people, or people already suffering disease, these results may not apply to the general population. This meta-analysis was later repeated and extended by the same authors, with the new analysis published by the Cochrane Collaboration; confirming the previous results. These two publications are consistent with some previous meta-analyzes that also suggested that Vitamin E supplementation increased mortality, and that antioxidant supplements increased the risk of colon cancer. However, the results of this meta-analysis are inconsistent with other studies such as the SU.VI.MAX trial, which suggested that antioxidants have no effect on cause-all mortality. Overall, the large number of clinical trials carried out on antioxidant supplements suggest that either these products have no effect on health, or that they cause a small increase in mortality in elderly or vulnerable populations.

While antioxidant supplementation is widely used in attempts to prevent the development of cancer, it has been proposed that antioxidants may, paradoxically, interfere with cancer treatments. This was thought to occur since the environment of cancer cells causes high levels of oxidative stress, making these cells more susceptible to the further oxidative stress induced by treatments.

As a result, by reducing the redox stress in cancer cells, antioxidant supplements could decrease the effectiveness of radiotherapy and chemotherapy. On the other hand, other reviews have suggested that antioxidants could reduce side effects or increase survival times.

Measurement and Levels in Food

Measurement of antioxidants is not a straightforward process, as this is a diverse group of compounds with different reactivities to different reactive oxygen species. In food science, the oxygen radical absorbance capacity (ORAC) has become the current industry standard for assessing antioxidant strength of whole foods, juices and food additives. Other measurement tests include the Folin-Ciocalteu reagent, and the Trolox equivalent antioxidant capacity assay.

Antioxidants are found in varying amounts in foods such as vegetables, fruits, grain cereals, eggs, meat, legumes and nuts. Some antioxidants such as lycopene and ascorbic acid can be destroyed by long-term storage or prolonged cooking. Other antioxidant compounds are more stable, such as the polyphenolic antioxidants in foods such as whole-wheat cereals and tea. The effects of cooking and food processing are complex, as these processes can also increase the bioavailability of antioxidants, such as some carotenoids in vegetables. In general, processed foods contain fewer antioxidants than fresh and uncooked foods, since the preparation processes may expose the food to oxygen.

Other antioxidants are not vitamins and are instead made in the body. For example, ubiquinol (coenzyme Q) is poorly absorbed from the gut and is made in humans through the mevalonate pathway. Another example is glutathione, which is made from amino acids. As any glutathione in the gut is broken down to free cysteine, glycine and glutamic acid before being absorbed, even large oral doses have little effect on the concentration of glutathione in the body. Although large amounts of sulfur-containing amino acids such as acetylcysteine can increase glutathione, no evidence exists that eating high levels of these glutathione precursors is beneficial for healthy adults. Supplying more of these precursors may be useful as part of the treatment of some diseases, such as acute respiratory distress syndrome, protein-energy malnutrition, or preventing the liver damage produced by paracetamol overdose.

Other compounds in the diet can alter the levels of antioxidants by acting as pro-oxidants. Here, consuming the compound causes oxidative stress, which the body responds to by inducing higher levels of antioxidant defenses such as antioxidant enzymes. Some of these compounds, such as isothiocyanates and curcumin, may be chemopreventive agents that either block the transformation of abnormal cells into cancerous cells, or even kill existing cancer cells.

Uses in Technology

Food Preservatives

Antioxidants are used as food additives to help guard against food deterioration. Exposure to oxygen and sunlight are the two main factors in the oxidation of food, so food is preserved by keeping in the dark and sealing it in containers or even coating it in wax, as with cucumbers. However, as oxygen is also important for plant respiration, storing plant materials in anaerobic conditions produces unpleasant flavours and unappealing colours. Consequently, packaging of fresh fruits and vegetables contains an ~8% oxygen atmosphere. Antioxidants are an especially important class of preservatives as, unlike bacterial or fungal spoilage, oxidation reactions still occur relatively rapidly in frozen or refrigerated food.

These preservatives include natural antioxidants such as ascorbic acid (AA, E300) and tocopherols (E306), as well as synthetic antioxidants such as propyl gallate (PG, E310), tertiary butylhydroquinone (TBHQ), butylated hydroxyanisole (BHA, E320) and butylated hydroxytoluene (BHT, E321). The most common molecules attacked by oxidation are unsaturated fats; oxidation causes them to turn rancid. Since oxidized lipids are often discolored and usually have unpleasant tastes such as metallic or sulfurous flavours, it is important to avoid oxidation in fat-rich foods. Thus, these foods are rarely preserved by drying; instead, they are preserved by smoking, salting or fermenting. Even less fatty foods such as fruits are sprayed with sulfurous antioxidants prior to air drying. Oxidation is often catalysed by metals, which is why fats such as butter should never be wrapped in aluminium foil or kept in metal containers. Some fatty foods such as olive oil are partially protected from oxidation by their natural content of antioxidants, but remain sensitive to photooxidation. Antioxidant preservatives are also added to fat-based cosmetics such as lipstick and moisturizers to prevent rancidity.

Phytochemical

Phytochemicals are chemical compounds that occur naturally in plants, such as beta-carotene. The term is generally used to refer to those chemicals that *may* affect health, but are not yet established as essential nutrients. While there is abundant scientific and government support for recommending diets rich in fruits and vegetables, there is only limited evidence that health benefits are due to specific phytochemicals.

Phytochemicals as Candidate Therapeutics

Phytochemicals have been used as drugs for millennia. For example, Hippocrates may have prescribed willow tree leaves to abate fever. Salicin, having anti-inflammatory and pain-relieving properties, was originally extracted from the bark of the white willow tree and later synthetically produced became the staple over-the-counter drug called Aspirin. There is evidence from laboratory studies that phytochemicals in fruits and vegetables may reduce the risk of cancer, possibly due to dietary fibers, polyphenol antioxidants and anti-inflammatory effects. Specific phytochemicals, such as fermentable dietary fibers, are allowed limited health claims by the US Food and Drug Administration (FDA).

An important cancer drug, Taxol (paclitaxel), is a phytochemical initially extracted and purified from the Pacific yew tree. Among phytochemicals from edible plants with promise for deterring disease, diindolylmethane, from Brassica vegetables (broccoli, cauliflower, cabbage, kale, Brussels sprouts) is being tested against recurring respiratory papillomatosis tumours (caused by the human papilloma virus), is in Phase III clinical trials for cervical dysplasia (a precancerous condition caused by the human papilloma virus) and is in several clinical trials for prostate cancer.

Some phytochemicals with physiological properties may be elements rather than complex organic molecules. Abundant in many fruits and vegetables, selenium, for example, is involved with major metabolic pathways, including thyroid hormone metabolism and immune function. Particularly, it is an essential nutrient and cofactor for the enzymatic synthesis of glutathione, an endogenous antioxidant.

Clinical Trials and Health Claim Status

There are currently many phytochemicals possibly having medicinal properties in clinical trials for a variety of diseases. Lycopene, for example, from tomatoes has been tested in clinical trials for cardiovascular diseases and prostate cancer. These studies, however, did not attain sufficient scientific agreement to conclude an effect on any disease. The FDA position reads:

> *"Very limited and preliminary scientific research suggests that eating one-half to one cup of tomatoes and/or tomato sauce a week may reduce the risk of prostate cancer. The United States Food and Drug Administration concludes that there is little scientific evidence supporting this claim."*

Likewise, although lutein and zeaxanthin may affect visual performance and inhibit macular degeneration and cataracts, there was insufficient scientific evidence from clinical trials for such a specific effect or health claim.

Food Processing and Phytochemicals

Phytochemicals in freshly harvested plant foods may be destroyed or removed by modern processing techniques, possibly including cooking. For this reason, industrially processed foods likely contain fewer phytochemicals and may thus be less beneficial than unprocessed foods. Absence or deficiency of phytochemicals in processed foods may contribute to increased risk of preventable diseases.

Interestingly, a converse example may exist in which lycopene, a phytochemical present in tomatoes, is either unchanged in content or made more concentrated by processing to juice or paste, maintaining good levels for bioavailability.

Secondary Metabolite

Secondary metabolites are organic compounds that are not directly involved in the normal growth, development, or reproduction of organisms. Unlike primary metabolites, absence of secondary metabolities does not result in immediate death, but rather in long-term impairment of the organism's survivability, fecundity, or aesthetics, or perhaps in no significant change at all. Secondary metabolites are often restricted to a narrow set of species within a phylogenetic group. Secondary metabolites often play an important role in plant defence against herbivory and other interspecies defenses.

Categories

Most of the secondary metabolites of interest to humankind fit into categories which classify secondary metabolites based on their biosynthetic origin.

Since secondary metabolites are often created by modified primary metabolite synthases, or "borrow" substrates of primary metabolite origin, these categories should not be interpreted as saying that all molecules in the category are secondary metabolites (for example the steroid category), but rather that there are secondary metabolites in these categories.

Small "Small Molecules"

- Alkaloids (usually a small, heavily derivatized amino acid):
  - Hyoscyamine, present in *Datura stramonium*
  - Atropine, present in *Atropa belladonna*, Deadly nightshade
  - Cocaine, present in *Erythroxylon coca* the Coca plant
  - Scopolamine, present in the *Solanaceae* (nightshade) plant family
  - Codeine and Morphine, present in *Papaver somniferum*, the opium poppy
  - Tetrodotoxin, a microbial product in Fugu and some salamanders
  - Vincristine & Vinblastine, mitotic inhibitors found in the Rosy Periwinkle.
- Terpenoids (come from semiterpene oligomerization):
  - Azadirachtin, (Neem tree)
  - Artemisinin, present in *Artemisia annua* Chinese wormwood
  - tetrahydrocannabinol, present in *cannabis sativa*
  - Steroids (Terpenes with a particular ring structure)
    - Saponins (plant steroids, often glycosylated)
- Glycosides (heavily modified sugar molecules):
  - Nojirimycin
  - Glucosinolates.

- Phenols:
  - o Resveratrol.
- Phenazines:
  - o Pyocyanin
  - o Phenazine-1-carboxylic acid (and derivatives)

Big "Small Molecules", Produced by Large, Modular, "Molecular Factories"

- Polyketides:
  - o Erythromycin
  - o Discodermolide.
- Fatty acid synthase products :
  - o FR-900848
  - o U-106305
  - o phloroglucinols.
- Nonribosomal peptides:
  - o Vancomycin
  - o Thiostrepton
  - o Ramoplanin
  - o Teicoplanin
  - o Gramicidin
  - o Bacitracin.
- Hybrids of the above three:
  - o Epothilone.
- Polyphenols.

Chapter

7

Lipid Metabolism

Lipids are a broad group of naturally occurring molecules which includes fats, waxes, sterols, fat-soluble vitamins (such as vitamins A, D, E and K), monoglycerides, diglycerides, phospholipids, and others. The main biological functions of lipids include energy storage, as structural components of cell membranes, and as important signalling molecules. Lipids may be broadly defined as hydrophobic or amphiphilic small molecules; the amphiphilic nature of some lipids allows them to form structures such as vesicles, liposomes, or membranes in an aqueous environment. Biological lipids originate entirely or in part from two distinct types of biochemical subunits or "building blocks": ketoacyl and isoprene groups.

Using this approach, lipids may be divided into eight categories: fatty acyls, glycerolipids, glycerophospholipids, sphingolipids, saccharolipids and polyketides (derived from condensation of ketoacyl subunits); and sterol lipids and prenol lipids (derived from condensation of isoprene subunits). Although the term *lipid* is sometimes used as a synonym for fats, fats are a subgroup of lipids called triglycerides. Lipids also encompass molecules such as fatty acids and their derivatives (including tri-, di-, and monoglycerides and phospholipids), as well as other sterol-containing metabolites such as cholesterol. Although humans and other mammals use various biosynthetic pathways to both break down and synthesize lipids, some essential lipids cannot be made this way and must be obtained from the diet.

Categories of Lipids

Fatty Acyls

Fatty acyls, a generic term for describing fatty acids, their conjugates and derivatives, are a diverse group of molecules synthesized by chain-elongation of an acetyl-CoA primer with malonyl-CoA or methylmalonyl-CoA groups in a process called fatty acid synthesis. They are made of a hydrocarbon chain that terminates with a carboxylic acid group; this arrangement confers the molecule with a polar, hydrophilic end, and a nonpolar, hydrophobic end that is insoluble in water. The fatty acid structure is one of the most fundamental categories of biological lipids, and is commonly used as a building block of more structurally complex lipids.

The carbon chain, typically between four to 24 carbons long, may be saturated or unsaturated, and may be attached to functional groups containing oxygen, halogens, nitrogen and sulfur. Where a double bond exists, there is the possibility of either a *cis* or *trans* geometric isomerism, which significantly affects the molecule's molecular configuration. *Cis*-double bonds cause the fatty acid chain to bend, an effect that is more pronounced the more double bonds there are in a chain. This in turn plays an important role in the structure and function of cell membranes. Most naturally occurring fatty acids are of the *cis* configuration, although the *trans* form does exist in some natural and partially hydrogenated fats and oils.

Examples of biologically important fatty acids are the eicosanoids, derived primarily from arachidonic acid and eicosapentaenoic acid, which include prostaglandins, leukotrienes, and thromboxanes. Other major lipid classes in the fatty acid category are the fatty esters and fatty amides. Fatty esters include important biochemical intermediates such as wax esters, fatty acid thioester coenzyme A derivatives, fatty acid thioester ACP derivatives and fatty acid carnitines. The fatty amides include N-acyl ethanolamines, such as the cannabinoid neurotransmitter anandamide.

Glycerolipids

Glycerolipids are composed mainly of mono-, di- and tri-substituted glycerols, the most well-known being the fatty acid esters of glycerol (triacylglycerols), also known as triglycerides. In these compounds, the three hydroxyl groups of glycerol are each esterified, usually by different fatty acids. Because they function as a food store, these lipids comprise the bulk of storage fat in animal tissues. The hydrolysis of the ester bonds of triacylglycerols and the release of glycerol and fatty acids from adipose tissue is called fat mobilization. Additional subclasses of glycerolipids are represented by glycosylglycerols, which are characterized by the presence of one or more sugar residues attached to glycerol via a glycosidic linkage. Examples of structures in this category are the digalactosyldiacylglycerols found in plant membranes and seminolipid from mammalian sperm cells.

Glycerophospholipids

Glycerophospholipids, also referred to as phospholipids, are ubiquitous in nature and are key components of the lipid bilayer of cells, as well as being involved in metabolism and cell signalling. Neural tissue (including the brain) contains relatively high amounts of glycerophospholipids, and alterations in their composition has been implicated in various neurological disorders. Glycerophospholipids may be subdivided into distinct classes, based on the nature of the polar headgroup at the *sn*-3 position of the glycerol backbone in eukaryotes and eubacteria, or the *sn*-1 position in the case of archaebacteria.

Examples of glycerophospholipids found in biological membranes are phosphatidylcholine (also known as PC, GPCho or lecithin), phosphatidylethanolamine (PE or GPEtn) and phosphatidylserine (PS or GPSer). In addition to serving as a primary component of cellular membranes and binding sites for intra- and intercellular proteins, some glycerophospholipids in eukaryotic cells, such as phosphatidylinositols and phosphatidic acids are either precursors of, or are themselves, membrane-derived second messengers. Typically, one or both of these hydroxyl groups are acylated with long-chain fatty acids, but there are also alkyl-linked and 1Z-alkenyl-linked (plasmalogen) glycerophospholipids, as well as dialkylether variants in archaebacteria.

Sphingolipids

Sphingolipids are a complex family of compounds that share a common structural feature, a sphingoid base backbone that is synthesized *de novo* from the amino acid serine and a long-chain fatty acyl CoA, then converted into ceramides, phosphosphingolipids, glycosphingolipids and other compounds. The major sphingoid base of mammals is commonly referred to as sphingosine. Ceramides (N-acyl-sphingoid bases) are a major subclass of sphingoid base derivatives with an amide-linked fatty acid. The fatty acids are typically saturated or mono-unsaturated with chain lengths from 16 to 26 carbon atoms. The major phosphosphingolipids of mammals are sphingomyelins (ceramide phosphocholines), whereas insects contain mainly ceramide phosphoethanolamines and fungi have phytoceramide phosphoinositols and mannose-containing headgroups. The glycosphingolipids are a diverse family of molecules composed of one or more sugar residues linked via a glycosidic bond to the sphingoid base. Examples of these are the simple and complex glycosphingolipids such as cerebrosides and gangliosides.

Sterol Lipids

Sterol lipids, such as cholesterol and its derivatives, are an important component of membrane lipids, along with the glycerophospholipids and sphingomyelins. The steroids, all derived from the same fused four-ring core structure, have different biological roles as hormones and signalling molecules. The eighteen-carbon (C18) steroids include the estrogen family whereas the C19 steroids comprise the androgens such as testosterone and androsterone.

The C21 subclass includes the progestogens as well as the glucocorticoids and mineralocorticoids. The secosteroids, comprising various forms of vitamin D, are characterized by cleavage of the B ring of the core structure. Other examples of sterols are the bile acids and their conjugates, which in mammals are oxidized derivatives of cholesterol and are synthesized in the liver. The plant equivalents are the phytosterols, such as β-sitosterol, stigmasterol, and brassicasterol; the latter compound is also used as a biomarker for algal growth. The predominant sterol in fungal cell membranes is ergosterol.

Prenol Lipids

Prenol lipids are synthesized from the 5-carbon precursors isopentenyl diphosphate and dimethylallyl diphosphate that are produced mainly via the mevalonic acid (MVA) pathway. The simple isoprenoids (linear alcohols, diphosphates, etc.) are formed by the successive addition of C5 units, and are classified according to number of these terpene units. Structures containing greater than 40 carbons are known as polyterpenes.

Carotenoids are important simple isoprenoids that function as antioxidants and as precursors of vitamin A. Another biologically important class of molecules is exemplified by the quinones and hydroquinones, which contain an isoprenoid tail attached to a quinonoid core of non-isoprenoid origin. Vitamin E and vitamin K, as well as the ubiquinones, are examples of this class. Prokaryotes synthesize polyprenols (called bactoprenols) in which the terminal isoprenoid unit attached to oxygen remains unsaturated, whereas in animal polyprenols (dolichols) the terminal isoprenoid is reduced.

Saccharolipids

Saccharolipids describe compounds in which fatty acids are linked directly to a sugar backbone, forming structures that are compatible with membrane bilayers. In the saccharolipids, a monosaccharide substitutes for the glycerol backbone present in glycerolipids and glycerophospholipids. The most familiar saccharolipids are the acylated glucosamine precursors of the Lipid A component of the lipopolysaccharides in Gram-negative bacteria. Typical lipid A molecules are disaccharides of glucosamine, which are derivatized with as many as seven fatty-acyl chains. The minimal lipopolysaccharide required for growth in *E. coli* is Kdo_2-Lipid A, a hexa-acylated disaccharide of glucosamine that is glycosylated with two 3-deoxy-D-manno-octulosonic acid (Kdo) residues.

Polyketides

Polyketides are synthesized by polymerization of acetyl and propionyl subunits by classic enzymes as well as iterative and multimodular enzymes that share mechanistic features with the fatty acid synthases. They comprise a large number of secondary metabolites and natural products from animal, plant, bacterial, fungal and marine sources, and have great structural diversity. Many polyketides are cyclic molecules whose backbones are often further modified by glycosylation, methylation, hydroxylation, oxidation, and/or other processes. Many commonly used anti-microbial, anti-parasitic, and anti-cancer agents are polyketides or polyketide derivatives, such as erythromycins, tetracyclines, avermectins, and antitumor epothilones.

Metabolism

The major dietary lipids for humans and other animals are animal and plant triglycerides, sterols, and membrane phospholipids. The process of lipid metabolism synthesizes and degrades the lipid stores and produces the structural and functional lipids characteristic of individual tissues.

Biosynthesis

In animals, when there is an oversupply of dietary carbohydrate, the excess carbohydrate is converted to triacylglycerol. This involves the synthesis of fatty acids from acetyl-CoA and the esterification of fatty acids in the production of triacylglycerol, a process called lipogenesis. Fatty acids are made by fatty acid synthases that polymerize and then reduce acetyl-CoA units. The acyl chains in the fatty acids are extended by a cycle of reactions that add the acetyl group, reduce it to an alcohol, dehydrate it to an alkene group and then reduce it again to an alkane group. The enzymes of fatty acid biosynthesis are divided into two groups, in animals and fungi all these fatty acid synthase reactions are carried out by a single multifunctional protein, while in plant plastids and bacteria separate enzymes perform each step in the pathway. The fatty acids may be subsequently converted to triacylglycerols that are packaged in lipoproteins and secreted from the liver.

The synthesis of unsaturated fatty acids involves a desaturation reaction, whereby a double bond is introduced into the fatty acyl chain. For example, in humans, the desaturation of stearic acid by stearoyl-CoA desaturase-1 produces oleic acid. The doubly unsaturated fatty acid linoleic acid as well as the triply unsaturated α-linolenic acid cannot be synthesized in mammalian tissues, and are therefore essential fatty acids and must be obtained from the diet.

Triacylglycerol synthesis takes place in the endoplasmic reticulum by metabolic pathways in which acyl groups in fatty acyl-CoAs are transferred to the hydroxyl groups of glycerol-3-phosphate and diacylglycerol.

Terpenes and isoprenoids, including the carotenoids, are made by the assembly and modification of isoprene units donated from the reactive precursors isopentenyl pyrophosphate and dimethylallyl pyrophosphate. These precursors can be made in different ways. In animals and archaea, the mevalonate pathway produces these compounds from acetyl-CoA, while in plants and bacteria the non-mevalonate pathway uses pyruvate and glyceraldehyde 3-phosphate as substrates. One important reaction that uses these activated isoprene donors is steroid biosynthesis. Here, the isoprene units are joined together to make squalene and then folded up and formed into a set of rings to make lanosterol. Lanosterol can then be converted into other steroids such as cholesterol and ergosterol.

Degradation

Beta oxidation is the metabolic process by which fatty acids are broken down in the mitochondria and/or in peroxisomes to generate acetyl-CoA. For the most part, fatty acids are oxidized by a mechanism that is similar to, but not identical with, a reversal of the process of fatty acid synthesis. That is, two-carbon fragments are removed sequentially from the carboxyl end of the acid after steps of dehydrogenation, hydration, and oxidation to form a beta-keto acid, which is split by thiolysis. The acetyl-CoA is then ultimately converted into ATP, CO_2, and H_2O using the citric acid cycle and the electron transport chain.

Hence the Krebs Cycle can start at acetyl-CoA when fat is being broken down for energy if there is little or no glucose available. The energy yield of the complete oxidation of the fatty acid palmitate is 106 ATP. Unsaturated and odd-chain fatty acids require additional enzymatic steps for degradation.

Nutrition and Health

Most of the lipid found in food is in the form of triacylglycerols, cholesterol and phospholipids. A minimum amount of dietary fat is necessary to facilitate absorption of fat-soluble vitamins (A, D, E and K) and carotenoids. Humans and other mammals have a dietary requirement for certain essential fatty acids, such as linoleic acid (an omega-6 fatty acid) and alpha-linolenic acid (an omega-3 fatty acid) because they cannot be synthesized from simple precursors in the diet. Both of these fatty acids are 18-carbon polyunsaturated fatty acids differing in the number and position of the double bonds. Most vegetable oils are rich in linoleic acid (safflower, sunflower, and corn oils). Alpha-linolenic acid is found in the green leaves of plants, and in selected seeds, nuts and legumes (particularly flax, rapeseed, walnut and soy). Fish oils are particularly rich in the longer-chain omega-3 fatty acids eicosapentaenoic acid (EPA) and docosahexaenoic acid (DHA). A large number of studies have shown positive health benefits associated with consumption of omega-3 fatty acids on infant development, cancer, cardiovascular diseases, and various mental illnesses, such as depression, attention-deficit hyperactivity disorder, and dementia. In contrast, it is now well-established that consumption of trans fats, such as those present in partially hydrogenated vegetable oils, are a risk factor for cardiovascular disease.

A few studies have suggested that total dietary fat intake is linked to an increased risk of obesity and diabetes. However, a number of very large studies, including the Women's Health Initiative Dietary Modification Trial, an eight year study of 49,000 women, the Nurses' Health Study and the Health Professionals Follow-up Study, revealed no such links. None of these studies suggested any connection between percentage of calories from fat and risk of cancer, heart disease or weight gain. The Nutrition Source, a website maintained by the Department of Nutrition at the Harvard School of Public Health, summarizes the current evidence on the impact of dietary fat: "Detailed research—much of it done at Harvard—shows that the total amount of fat in the diet isn't really linked with weight or disease."

Fatty Acid Metabolism

Fatty acids are an important source of energy and adenosine triphosphate (ATP} for many cellular organisms. Excess fatty acids, glucose, and other nutrients can be stored efficiently as fat. Triglycerides yield more than twice as much energy for the same mass as do carbohydrates or proteins. All cell membranes are built up of phospholipids, each of which contains two fatty acids. Fatty acids are also used for protein modification. The metabolism of fatty acids, therefore, consists of catabolic processes that generate energy and primary metabolites from fatty acids, and anabolic processes that create biologically important molecules from fatty acids and other dietary carbon sources.

Overview

- Lipolysis is carried out by lipases.
- Once freed from glycerol, free fatty acids can enter blood and muscle fiber by diffusion.
- Beta oxidation splits long carbon chains of the fatty acid into acetyl CoA, which can eventually enter the TCA cycle.

Briefly, β-oxidation or lipolysis of free fatty acids is as follows:

1. Dehydrogenation by acyl-CoA dehydrogenase, yielding 1 $FADH_2$
2. Hydration by enoyl-CoA hydratase
3. Dehydrogenation by 3-hydroxyacyl-CoA dehydrogenase, yielding 1 NADH
4. Cleavage by thiolase, yielding 1 acetyl-CoA and a fatty acid that has now been shortened by 2 carbons (acyl-CoA)

This cycle repeats until the FFA has been completely reduced to acetyl-CoA or, in the case of fatty acids with odd numbers of carbon atoms, acetyl-CoA and 1 mol of propionyl-CoA per mol of fatty acid.

Fatty Acids as an Energy Source

Fatty acids, stored as triglycerides in an organism, are an important source of energy because they are both reduced and anhydrous. The energy yield from a gram of fatty acids is approximately 9 Kcal (37 kJ), compared to 4 Kcal/g (17 kJ/g) for carbohydrates. Since the hydrocarbon portion of fatty acids is hydrophobic, these molecules can be stored in a relatively anhydrous (water-free) environment. Carbohydrates, on the other hand, are more highly hydrated.

For example, 1 g of glycogen can bind approximately 2 g of water, which translates to 1.33 Kcal/g (4 Kcal/3 g). This means that fatty acids can hold more than six times the amount of energy per unit of storage mass. Put another way, if the human body relied on carbohydrates to store energy, then a person would need to carry 67.5 lb (31 kg) of hydrated glycogen to have the energy equivalent to 10 lb (5 kg) of fat. Hibernating animals provide a good example for utilizing fat reserves as fuel. For example, bears hibernate for about 7 months, and, during this entire period, the energy is derived from degradation of fat stores.

Digestion and Transport

Fatty acids are usually ingested as triglycerides, which cannot be absorbed by the intestine. They are broken down into free fatty acids and monoglycerides by pancreatic lipase, which forms a 1:1 complex with a protein called colipase, which is necessary for its activity. The activated complex can work only at a water-fat interface. Therefore, it is essential that fatty acids (FA) be emulsified by bile salts for optimal activity of these enzymes. People having had their gallbladder removed due to gall stones have, as a consequence, great difficulty digesting fats.

The digestion products of triglycerides are absorbed primarily as free fatty acids and 2-monoglycerides, but a small fraction are absorbed as free glycerol and as diglycerides. Once across the intestinal barrier, they are reformed into triglycerides and packaged into chylomicrons or liposomes, which are released into the lacteals, the capillaries of the lymph system and then into the blood. Eventually, they bind to the membranes of hepatocytes, adipocytes or muscle fibers, where they are either stored or oxidized for energy.

The liver acts as a major organ for fatty acid treatment, processing chylomicron remnants and liposomes into the various lipoprotein forms, in particular VLDL and LDL. Fatty acids synthesized by the liver are converted to triglyceride and transported to the blood as VLDL. In peripheral tissues, lipoprotein lipase digests part of the VLDL into LDL and free fatty acids, which are taken up for metabolism. This is done by the removal of the triglycerides contained in the

VLDL. What is left of the VLDL absorbs cholesterol from other circulating lipoproteins, becoming LDLs. LDL is absorbed via LDL receptors. This provides a mechanism for absorption of LDL into the cell, and for its conversion into free fatty acids, cholesterol, and other components of LDL. The liver controls the concentration of cholesterol in the blood by removing LDL. Another type of lipoprotein known as high-density lipoprotein, or HDL collects cholesterol, glycerol and fatty acids from the blood and transports them to the liver. In summary:

- Chylomicrons carry diet-derived lipids to body cells
- VLDLs carry lipids synthesized by the liver to body cells
- LDLs carry cholesterol around the body
- HDLs carry cholesterol from the body back to the liver for breakdown and excretion.

When blood sugar is low, glucagon signals the adipocytes to activate hormone-sensitive lipase, and to convert triglycerides into free fatty acids. These have very low solubility in the blood, typically about 1 μM. However, the most abundant protein in blood, serum albumin, binds free fatty acids, increasing their effective solubility to ~ 1 mM. Thus, serum albumin transports fatty acids to organs such as muscle and liver for oxidation when blood sugar is low.

Transport and Oxidation

The neutral lipids stored in adipocytes (and in steroid synthesizing cells of the adrenal cortex, ovary, and testes) in the form of lipid droplets, with a core of sterol esters and triacylglycerols surrounded by a monolayer of phospholipids are coated with Perilipin, a protein that acts as a protective coating from the body's natural lipases, such as hormone-sensitive lipase,. However, when a hormone such as epinepherine and glucagon are secreted in response to low levels of glucose, this triggers an intracellular secondary messenger cascade which phosphorylates hormone-sensitive lipase to break triglycerides into glycerol and free fatty acids for use in metabolism, a process called lipolysis.

The free fatty acids the move into the blood stream where they are bound by serum albumin and transported to the tissue needing fuel. Once the fatty acids reach the target tissue, they are released by serum albumin and cross into the cytosol. The enzymes used in fatty acid oxidation in animal cells are located in the mitochondrial matrix (as was demonstrated by Eugene P. Kennedy and Albert Lehninger in 1948).

Free fatty acid chains of more than 12 carbons require the help of membrane transporters to cross into the membrane into the mitochondria, where they undergo Fatty acid degredation.

Fatty acid degradation is the process in which fatty acids are broken down, resulting in release of energy. It includes three major steps:

- Activation and transport into the mitochondria
- β-Oxidation
- Electron transport chain.

Fatty acids are transported across the outer mitochondrial membrane by carnitine-palmitoyl transferase I (CPT-I), and then couriered across the inner mitochondrial membrane by carnitine. Once inside the mitochondrial matrix, fatty acyl-carnitine reacts with coenzyme A to release the fatty acid and produce acetyl-CoA. CPT-I is believed to be the rate limiting step in fatty acid oxidation.

Once inside the mitochondrial matrix, fatty acids undergo β-oxidation. During this process, two-carbon molecules acetyl-CoA are repeatedly cleaved from the fatty acid. Acetyl-CoA can then enter the TCA cycle, which produces NADH and $FADH_2$. NADH and $FADH_2$ are subsequently used in the electron transport chain to produce ATP, the energy currency of the cell.

Besides β-oxidation, other oxidative pathways are sometimes employed. α-Oxidation is used for branched fatty acids that cannot directly undergo β-oxidation. The smooth ER of the liver can perform ω-oxidation, which is primarily for detoxification but can become much more prevalent in cases of defective β-oxidation. Fatty acids with very long chains (20 or more carbons) are first broken down to a manageable size in peroxisomes.

Synthesis

Fatty acid synthesis is the creating of fatty acids from acetyl-CoA and malonyl-CoA precursors through action of enzymes called fatty acid synthases. It is an important part of the lipogenesis process, which-together with glycolysis-stands behind creating fats from blood sugar in living organisms.

De Novo Synthesis in Humans

In humans, fatty acids are predominantly formed in the liver and lactating mammary glands, and, to a lesser extent, the adipose tissue. Most acetyl-CoA is formed from pyruvate by pyruvate dehydrogenase in the mitochondria. Acetyl-CoA produced in the mitochondria is condensed with oxaloacetate by citrate synthase to form citrate, which is then transported into the cytosol and broken down to yield acetyl-CoA and oxaloacetate by ATP citrate lyase. Oxaloacetate in the cytosol is reduced to malate by cytoplasmic malate dehydrogenase, and malate is transported back into the mitochondria to participate in the Citric acid cycle.

Elongation

Much like β-oxidation, elongation occurs via four recurring reactions shown below. In these diagrams, the acetyl and malonyl units are shown as their Acyl carrier protein (ACP) thioesters: This is how fatty acids are synthesized in microorganisms and plants. However, in animals, these same reactions occur on a large dimeric protein, Fatty acid synthase, which has the full complement of enzymatic activities required to synthesize and liberate a free fatty acid.

The end-product of these reactions is always palmitate. No intermediates are released until palmitate is formed. After release from the ACP, palmitate is esterified to Coenzyme A, as this is done with all free fatty acids inside cells to prevent lysis of the cell membranes. If further elongation has to happen, the palmitoyl CoA, or any other acyl-CoA from a dietary acid, then moves into the endoplasmic reticulum, where it can be elongated up to a length of 20 to 24 carbons by the same chain of reactions as happens in the cytosolic ACP.

Regulation and Control

It has long been held that hormone-sensitive lipase (HSL) is the enzyme that hydrolyses triacylglycerides to free fatty acids from fats (lipolysis). However, more recently it has been shown that at most HSL converts triacylglycerides to monoglycerides and free fatty acids. Monoglycerides are hydrolysed by monoglyceride lipase; adipose triglyceride lipase may have a special role in converting triacylglycerides to diacylglycerides, while diacylglycerides are the best substrate for HSL.. HSL is regulated by the hormones insulin, glucagon, norepinephrine, and epinephrine.

Glucagon is associated with low blood glucose, and epinephrine is associated with increased metabolic demands. In both situations, energy is needed, and the oxidation of fatty acids is increased to meet that need. Glucagon, norepinephrine, and epinephrine bind to G protein-coupled receptors that activate adenylate cyclase to produce cyclic AMP. As a consequence, cAMP activates protein kinase A, which phosphorylates (and activates) hormone-sensitive lipase. When blood glucose is high, lipolysis is inhibited by insulin. Insulin activates protein phosphatase 2A, which dephosphorylates HSL, thereby inhibiting its activity. Insulin also activates the enzyme phosphodiesterase, which breaks down cAMP and stops the re-phosphorylation effects of protein kinase A.

Steroidogenesis

Steroidogenesis is the biological process by which steroids are generated from cholesterol and transformed into other steroids. The pathways of steroidogenesis differ between different species, but the pathways of human steroidogenesis.

Products of steroidogenesis include:

- androgens
    - testosterone
- estrogens and progesterone
- corticoids
    - cortisol
    - aldosterone.

Lipolysis

Lipolysis is the breakdown of lipids. Metabolically it is the hydrolysis of triglycerides into free fatty acids followed by further degradation, into acetyl units, by beta oxidation. Ketones are produced, and are found in large quantities in ketosis (a state in metabolism occurring when the liver converts fat into fatty acids and ketone bodies which can be used by the body for energy). Lipolysis testing strips such as Ketostix are used to recognize ketosis.

The following hormones induce lipolysis: epinephrine, norepinephrine, glucagon, growth hormone and cortisol (though cortisol's actions are still unclear). These trigger 7TM receptors (G protein-coupled receptors), which activate adenylate cyclase. This results in increased production of cAMP, which activates protein kinase A, which subsequently activate lipases found in adipose tissue.

Triglycerides are transported through the blood to appropriate tissues (adipose, muscle, etc) by lipoproteins such as chylomicrons. Triglycerides present on the chylomicrons undergo lipolysis by the cellular lipases of target tissues which yields glycerol and free fatty acids. Free fatty acids released into the blood are then available for cellular uptake. Free fatty acids not immediately taken up by cells may bind to albumin for transport to surrounding tissues that require energy. Serum Albumin is the major carrier of free fatty acids in the blood. The glycerol also enters the bloodstream and is absorbed by the liver or kidney where it is converted to glycerol 3-phosphate by the enzyme glycerol kinase.

Hepatic glycerol 3-phosphate is mostly converted into dihydroxyacetonephosphate (DHAP) and then glyceraldehyde 3-phosphate (GA3P) to rejoin the glycolysis and gluconeogenesis pathway. While lipolysis is triglyceride hydrolysis, the process by which triglycerides are broken down, esterification is the process by which triglycerides are formed. Esterification and lipolysis are essentially reversals of one another. Lipolysis during stress occurs in the fat cells which in turn increases cholestrol during chronic enduring stress.

Lipogenesis

Lipogenesis is the process by which acetyl-CoA is converted to fats. The former is an intermediate stage in metabolism of simple sugars, such as glucose, a main source of energy of living organisms. Through lipogenesis, the energy can be efficiently stored in the form of fats. Lipogenesis encompasses the processes of fatty acid synthesis and subsequent triglyceride synthesis (when fatty acids are esterified with glycerol to form fats). The products are secreted from the liver in the form of very-low-density lipoproteins (VLDL).

Fatty Acid Synthesis

Fatty acids synthesis starts with acetyl-CoA and builds up by the addition of two carbon units. The synthesis occurs in the cytoplasm in contrast to the degradation (oxidation), which occurs in the mitochondria. Many of the enzymes for the fatty acid synthesis are organized into a multienzyme complex called fatty acid synthetase.

Control and Regulation

Insulin is an indicator of the blood sugar level of the body, as its concentration increases proportionally with blood sugar levels. Thus, a large insulin level is associated with the fed state. As one might expect, therefore, it increases the rate of storage pathways, such as lipogenesis. Insulin stimulates lipogenesis in two main ways: The enzymes pyruvate dehydrogenase (PDH), which forms acetyl-CoA, and acetyl-CoA carboxylase (ACC), which forms malonyl-CoA, are obvious control points. These are activated by insulin. So a high insulin level leads to an overall increase in the levels of malonyl-CoA, which is the substrate required for fatty acids synthesis.

PDH Dephosphorylation

Pyruvate dehydrogenase dephosphorylation is increased with the release of insulin. The dephosphorylated form is more active. This mechanism is not clear, as insulin binds to extra-cellular parts of protein receptors and PDH is found in the mitochondrial matrix.

There must be some sort of secondary messenger process. This mechanism leads to the increased rate of catalysis of this enzyme, so increases the levels of acetyl-CoA. Increased levels of acetyl-CoA will increase the flux through not only the fat synthesis pathway but also the citric acid cycle.

Acetyl-CoA Carboxylase

Insulin affects ACC in a similar way to PDH. It leads to its dephosphorylation. Glucagon has an antagonistic effect and increases phosphorylation, thereby inhibiting ACC and slowing fat synthesis.

This inhibition mechanism is thought to have something to do with ACC-dependent protein kinase. Affecting ACC affects the rate of acetyl-CoA conversion to malonyl-CoA. Increased malonyl-CoA level pushes the equilibrium over to increase production of fatty acids through biosynthesis.

AMP-activated protein kinase acts as a measure of the ATP needs of a cell and acts to phosphorylase ACC. When ATP is depleted, there is a rise in 5'AMP. This rise activates AMP-activated protein kinase, which phosphorylates ACC, thereby inhibits fat synthesis.

This is a useful way to ensure that glucose is not diverted down a storage pathway in times when energy levels are low. ACC is also activated by citrate. This means that, when there is abundant acetyl-CoA in the cell cytoplasm for fat synthesis, it proceeds at an appropriate rate.

Note: Research now shows that glucose metabolism (exact metabolite to be determined), aside from insulin's influence on lipogenic enzyme genes, can induce the gene products for liver's pyruvate kinase, acetyl-CoA carboxylase, and fatty acid synthase. These genes are induced by the transcription factors ChREBP/Mlx via high blood glucose levels and presently unknown signalling events. Insulin induction is due to SREBP-1c, which is also involved in cholesterol metabolism.

Storage and Expression of Genetic Information

The human genome (or genetic material) is comprised of deoxyribonucleic acid (DNA) that encodes information required for all life processes, including growth, development, reproduction, and even cell death. The functional units within the genome are called genes. Genes are hereditary regions of DNA that encode functional molecules, either proteins or ribonucleic acid (RNA) species. The human genome encodes approximately 100,000 genes on 23 chromosomes. DNA resides in a specific compartment within the cell, known as the nucleus. Each nucleated human cell within an individual, regardless of its origin, contains identical DNA.

However, the genetic code is expressed or read differently in each cell type. Gene expression refers to the processes in which the genetic code is deciphered to produce a functional macromolecule, either protein or RNA. While some genes are expressed in all cells, others are expressed exclusively in certain tissues or organs. This selective reading of the code imparts very different chemical, functional, and morphological properties to each cell type and ultimately defines the function of a tissue or organ.

Genes can also display temporal specific expression. For example, some genes are expressed only in the fetus, while other genes are not expressed until puberty or adulthood. Therefore, human DNA not only contains all of the genes required to assemble a human organism, but also encodes information that directs where, when, and how much an individual gene will be expressed.

Mechanisms of Gene Expression

Genes encode proteins. Proteins are polymers of amino acid building blocks that serve a variety of biological functions. Proteins can function as intracellular scaffolds that maintain cell integrity; others are transporters that permit specific nutrients and other small molecules to enter the cell. Proteins also can be enzymes that catalyse the many chemical reactions required for cell survival. While DNA is present in the nucleus, protein synthesis occurs in the cytoplasm, a separate compartment within the cell.

Therefore, an intermediate molecule is needed to transfer the genetic information from the nuclear compartment that contains the code to the cytoplasmic compartment where the code is read. This intermediate molecule is termed "messenger RNA," and it is a short-lived functional copy of the genetic code. The process by which the genetic code, DNA, is copied to make a messenger RNA molecule occurs in the nucleus and is termed "transcription." The process of reading the genetic code from a messenger RNA molecule occurs in the cytoplasm and is termed "translation." The end product of translation is a protein molecule.

The expression of some genes is predetermined and cannot be altered. However, the expression of other genes, particularly those involved in nutrient storage, processing, and metabolism, is dynamic, and can be influenced by the cell's environment. Therefore, in some instances, gene expression can be an adaptive process. It is now well established that the expression of many genes is, in part, constrained by the nutrient environment—giving credence to the old adage, "you are what you eat."

There are more than forty nutrients that are essential for mammals, and deficiencies in any of these nutrients have direct impacts on health. The cellular demand for these nutrients can vary as a function of growth, development, age, reproductive status, and immunity. However, for many organisms, the availability of nutrients can vary daily, weekly, and, in some cases, seasonally without notable changes in health.

This is because organisms adapt to nutrient supply by altering gene expression. This alteration, in turn, enables cells to increase their storage capacity for certain nutrients, alter the absorption or excretion of certain nutrients, use alternative metabolic pathways, or reprogram metabolic pathways.

Nutrient Control of RNA Synthesis

The best-understood signalling molecules are hormones such as estrogen and testosterone. A hormone is produced by a particular tissue and causes a specific biological change in the same tissue or a different tissue located elsewhere in the body. In some cases, these molecules can enter a cell and bind to a particular protein molecule, termed a "nuclear receptor." The receptor-hormone complex then travels to the nucleus, binds very specific regions of the DNA, and turns on the expression of genes not normally expressed in the absence of the hormone. This change in gene expression imparts new functional roles to individual cells, which can impact the entire organism greatly. In this manner, diverse biological processes can be initiated, including puberty or menstruation. Certain nutrients can also influence gene expression in a similar manner. Nutrients, including vitamin A, vitamin

D, and certain fatty acids, bind nuclear receptors and influence the expression of genes. These nutrient receptors enable cells to sense their nutrient environment and adjust cellular metabolism accordingly by altering the expression of genes.

Nutrient Control of Protein Synthesis

Nuclear receptors are effective in reprogramming DNA transcription to adapt to nutrient environments. Other mechanisms exist to alter gene expression without changing rates of DNA transcription. In fact, alteration of translation is a common mechanism that permits cells to adapt rapidly to changing nutrient environments. Iron is the paradigm for nutrient regulation of gene expression at the level of translation.

Iron is a critical component of many metabolic proteins and enzymes involved in oxygen transport, energy metabolism, and DNA synthesis. Iron deficiency results in several disease states, including anemia. Therefore, the body must retain sufficient iron stores to stave off such pathologies. However, iron is also a potent oxidant and, if not bound by proteins in the cell, it can destroy DNA and proteins, and catalyse events that initiate cancer. Therefore, the body must store iron, but in such a manner that prevents the iron from destroying the integrity of its cells.

Cells are protected from the deleterious effects of iron by sequestering it in a protein shell called ferritin. Cellular iron is stored in ferritin until required. Ferritin synthesis is rapidly induced when cells are exposed to iron and this increased synthesis is directly regulated by iron. Cells contain an iron-sensing protein called the iron regulatory protein (IRP). This protein binds either iron or ferritin messenger RNA but cannot bind both molecules simultaneously. When iron is not available to the cell, the intracellular concentration of nonprotein-bound iron is very low and IRP does not contain bound iron. This results in IRP being available to bind ferritin messenger RNA, which stops new ferritin synthesis. However, when cells are exposed to iron, IRP contains a bound iron molecule and cannot bind ferritin messenger RNA, and ferritin synthesis occurs. In this manner, the iron storage protein ferritin is only synthesized when it is required to store new iron.

Permanent Adaptation to Nutrient Supply

There is accumulating evidence that prenatal and postnatal nutrition can permanently alter cellular metabolism by altering gene expression throughout adulthood, a phenomenon termed "metabolic imprinting." Low birth weight, which occurred in infants born to survivors of the Dutch Famine of 1944–1945, has been linked to an increased risk of chronic disease later in life, including adult obesity, insulin resistance, hypertension, and cardiovascular disease.

The susceptibility to these disease states is influenced both by dietary habits as well as one's genetic predisposition or heritage. Although the biological basis for metabolic imprinting is not yet proven, the suggestion that gene expression can be programmed by fetal and postnatal nutrient environment has far-reaching implications. For many adult chronic disease states, dietary management is an important component of the therapy. If metabolic imprinting occurs, dietary management early in life may also be advantageous in preventing numerous chronic disease states that do not surface until adulthood.

Nutritional Modulation of Gene Expression in Health and Disease

The relationships between nutrient availability and adaptive changes in gene expression are critical to understanding the role of nutrition in health and disease. For many nutrients, either dietary insufficiency or excess can result in or contribute to disease onset. Nutrient modulation of gene expression serves to protect the cell from the deleterious effects of both under-nutrition and over-nutrition. Hereditary hyperferritinemia-cataract syndrome is a human disorder associated with altered regulation of iron homeostasis. Affected individuals have mutations in a ferritin gene that result in the synthesis of a ferritin messenger RNA that encodes a normal functional ferritin protein, but the mutation does not permit IRP to bind to the messenger RNA and stop translation. Therefore, these individuals can no longer regulate ferritin levels in response to changes in iron intake. As a result of this mutation, these individuals contract early-onset bilateral cataract associated with a progressive decrease in visual acuity. Ongoing research is identifying many other nutrient-related disease states that result from disregulation of nutrient control of gene expression.

Familial Hypercholesterolemia

Familial hypercholesterolemia is a disease that results from genetic mutations in the low-density lipoprotein (LDL) receptor. It is one of the most common inborn errors of metabolism. Individuals with one mutated copy of the gene (referred to as heterozygotes for this mutation) number about one in five hundred, whereas one in a million individuals carry two mutated copies of this gene (referred to as homozygotes for a mutation). LDL receptors are necessary for transporting LDL into cells from serum. LDL is a major cholesterol transport lipoprotein in human plasma, and individuals with LDL receptor mutations accumulate LDL in serum because LDL transport into cells is impaired. Plasma cholesterol levels from affected heterozygotes range from 350 to 550 mg/dl, and these values can exceed 1000 mg/dl for affected homozygotes. Individuals with elevated serum cholesterol have a high risk for developing heart disease at very young ages.

Cholesterol is an important component of cell membranes, and it serves to decrease their fluidity. Mammals can synthesize cholesterol in the absence of sufficient dietary cholesterol, but the endogenous biosynthesis is tightly regulated and inhibited by dietary cholesterol supply. A transcription factor known as SREBP (Sterol Response Element Binding Protein) regulates the expression of a gene that encodes a key enzyme that is necessary for cholesterol biosynthesis. When dietary intake of cholesterol is adequate, this transcription factor is sequestered in the membranes of the Golgi compartment of the cell and is inactive.

When cellular cholesterol levels fall, however, the membranes in the Golgi become more fluid. This results in the liberation of SREBP from the membrane and enables it to travel to the nucleus, where it can activate the expression of genes that are necessary for cholesterol biosynthesis. Through this regulation, cholesterol biosynthesis occurs only when dietary sources are limited. However, this mechanism of gene regulation fails in familial hypercholesterolemia. Because LDL cholesterol is not effectively transported into cells, the cells cannot sense extracellular cholesterol levels, and therefore cellular SREBP activity and cholesterol biosynthesis is constantly activated. Activated cholesterol biosynthesis serves to further increase serum LDL concentrations in affected individuals.

Chapter

8

Deoxyribonucleic Acid

Deoxyribonucleic acid, or DNA, is a nucleic acid that contains the genetic instructions used in the development and functioning of all known living organisms with the exception of some viruses. The main role of DNA molecules is the long-term storage of information. DNA is often compared to a set of blueprints, like a recipe or a code, since it contains the instructions needed to construct other components of cells, such as proteins and RNA molecules. The DNA segments that carry this genetic information are called genes, but other DNA sequences have structural purposes, or are involved in regulating the use of this genetic information.

DNA consists of two long polymers of simple units called nucleotides, with backbones made of sugars and phosphate groups joined by ester bonds. These two strands run in opposite directions to each other and are therefore anti-parallel. Attached to each sugar is one of four types of molecules called bases. It is the sequence of these four bases along the backbone that encodes information. This information is read using the genetic code, which specifies the sequence of the amino acids within proteins. The code is read by copying stretches of DNA into the related nucleic acid RNA, in a process called transcription.

Nucleic acids are polynucleotides, that is they are made up of chains of many repeating units called nucleotides. The DNA molecule actually consists of two polynucleotide chains that spiral around an imaginary axis to form a double helix (spiral.) Nucleic acid molecules are incredibly complex, containing the code that guarantees the accurate ordering of the 20 amino acids in all proteins made by living cells. Surprisingly though there are only a few different nucleotides. And, in fact, only four different nucleotide units comprise DNA, the nucleic acid of interest to the genealogist.

Within cells, DNA is organized into long structures called chromosomes. These chromosomes are duplicated before cells divide, in a process called DNA replication. Eukaryotic organisms (animals, plants, fungi, and protists) store most of their DNA inside the cell nucleus and some of their DNA in organelles, such as mitochondria or chloroplasts. In contrast, prokaryotes (bacteria and archaea) store their DNA only in the cytoplasm. Within the chromosomes, chromatin proteins such as histones compact and organize DNA. These compact structures guide the interactions between DNA and other proteins, helping control which parts of the DNA are transcribed.

Alternating sugar and phosphate units form the two sides of a ladder-shaped arrangement with the rungs or steps each formed by a pair of nucleotide bases (bps) The sides of the ladder are held together by weak electrostatic bonds between base pairs called hydrogen bonds (sort of the velcro of the molecular world.) The human genome has about 3,000,000,000 bps!

Properties

Chemical structure of DNA. Hydrogen bonds shown as dotted lines. DNA is a long polymer made from repeating units called nucleotides. As first discovered by James D. Watson and Francis Crick, the structure of DNA of all species comprises two helical chains each coiled round the same axis, and each with a pitch of 34 Ångströms (3.4 nanometres) and a radius of 10 Ångströms (1.0 nanometres). According to another study, when measured in a particular solution, the DNA chain measured 22 to 26 Ångströms wide (2.2 to 2.6 nanometres), and one nucleotide unit measured 3.3 Å (0.33 nm) long. Although each individual repeating unit is very small, DNA polymers can be very large molecules containing millions of nucleotides. For instance, the largest human chromosome, chromosome number 1, is approximately 220 million base pairs long.

In living organisms, DNA does not usually exist as a single molecule, but instead as a pair of molecules that are held tightly together. These two long strands entwine like vines, in the shape of a double helix. The nucleotide repeats contain both the segment of the backbone of the molecule, which holds the chain together, and a base, which interacts with the other DNA strand in the helix. A base linked to a sugar is called a nucleoside and a base linked to a sugar and one or more phosphate groups is called a nucleotide. If multiple nucleotides are linked together, as in DNA, this polymer is called a polynucleotide.

The backbone of the DNA strand is made from alternating phosphate and sugar residues. The sugar in DNA is 2-deoxyribose, which is a pentose (five-carbon) sugar. The sugars are joined together by phosphate groups that form phosphodiester bonds between the third and fifth carbon atoms of adjacent sugar rings. These asymmetric bonds mean a strand of DNA has a direction.

In a double helix the direction of the nucleotides in one strand is opposite to their direction in the other strand: the strands are *antiparallel*. The asymmetric ends of DNA strands are called the 52 (*five prime*) and 32 (*three prime*) ends, with the 5' end having a terminal phosphate group and the 3' end a terminal hydroxyl group. One major difference between DNA and RNA is the sugar, with the 2-deoxyribose in DNA being replaced by the alternative pentose sugar ribose in RNA.

The DNA double helix is stabilized by hydrogen bonds between the bases attached to the two strands. The four bases found in DNA are adenine (abbreviated A), cytosine (C), guanine (G) and thymine (T). These four bases are attached to the sugar/phosphate to form the complete nucleotide, as shown for adenosine monophosphate.

These bases are classified into two types; adenine and guanine are fused five- and six-membered heterocyclic compounds called purines, while cytosine and thymine are six-membered rings called pyrimidines. A fifth pyrimidine base, called uracil (U), usually takes the place of thymine in RNA and differs from thymine by lacking a methyl group on its ring. Uracil is not usually found in DNA, occurring only as a breakdown product of cytosine. In addition to RNA and DNA, a large number of artificial nucleic acid analogues have also been created to study the proprieties of nucleic acids, or for use in biotechnology.

Grooves

Twin helical strands form the DNA backbone. Another double helix may be found by tracing the spaces, or grooves, between the strands. These voids are adjacent to the base pairs and may provide a binding site. As the strands are not directly opposite each other, the grooves are unequally sized. One groove, the major groove, is 22 Å wide and the other, the minor groove, is 12 Å wide. The narrowness of the minor groove means that the edges of the bases are more accessible in the major groove.

As a result, proteins like transcription factors that can bind to specific sequences in double-stranded DNA usually make contacts to the sides of the bases exposed in the major groove. This situation varies in unusual conformations of DNA within the cell, but the major and minor grooves are always named to reflect the differences in size that would be seen if the DNA is twisted back into the ordinary B form.

Base Pair

Each type of base on one strand forms a bond with just one type of base on the other strand. This is called complementary base pairing. Here, purines form hydrogen bonds to pyrimidines, with A bonding only to T, and C bonding only to G. This arrangement of two nucleotides binding together across the double helix is called a base pair. As hydrogen bonds are not covalent, they can be broken and rejoined relatively easily. The two strands of DNA in a double helix can therefore be pulled apart like a zipper, either by a mechanical force or high temperature. As a result of this complementarity, all the information in the double-stranded sequence of a DNA helix is duplicated on each strand, which is vital in DNA replication. Indeed, this reversible and specific interaction between complementary base pairs is critical for all the functions of DNA in living organisms. Top, a GC base pair with three hydrogen bonds. Bottom, an AT base pair with two hydrogen bonds. Non-covalent hydrogen bonds between the pairs are shown as dashed lines.

The two types of base pairs form different numbers of hydrogen bonds, AT forming two hydrogen bonds, and GC forming three hydrogen bonds. DNA with high GC-content is more stable than DNA with low GC-content, but contrary to popular belief, this is not due to the extra hydrogen bond of a GC base pair but rather the contribution of stacking interactions (hydrogen bonding merely provides specificity of the pairing, not stability).

As a result, it is both the percentage of GC base pairs and the overall length of a DNA double helix that determine the strength of the association between the two strands of DNA. Long DNA helices with a high GC content have stronger-interacting strands, while short helices with high AT content have weaker-interacting strands. In biology, parts of the DNA double helix that need to separate easily, such as the TATAAT Pribnow box in some promoters, tend to have a high AT content, making the strands easier to pull apart.

In the laboratory, the strength of this interaction can be measured by finding the temperature required to break the hydrogen bonds, their melting temperature (also called T_m value). When all the base pairs in a DNA double helix melt, the strands separate and exist in solution as two entirely independent molecules. These single-stranded DNA molecules (*ssDNA*) have no single common shape, but some conformations are more stable than others.

The size of an individual gene or an organism's entire genome is often measured in base pairs because DNA is usually double-stranded. Hence, the number of total base pairs is equal to the number of nucleotides in one of the strands (with the exception of non-coding single-stranded regions of telomeres). The haploid human genome (23 chromosomes) is estimated to be about 3 billion base pairs long and to contain 20,000–25,000 distinct genes. A kilobase (kb) is a unit of measurement in molecular biology equal to 1000 base pairs of DNA or RNA.

Hydrogen Bonding and Stability

Hydrogen bonding is the chemical interaction that underlies the base-pairing rules. Appropriate geometrical correspondence of hydrogen bond donors and acceptors allows only the "right" pairs to form stably. DNA with high GC-content is more stable than DNA with low GC-content, but contrary to popular belief, the hydrogen bonds do not stabilize the DNA significantly and stabilization is mainly due to stacking interactions.

The larger nucleobases, adenine and guanine, are members of a class of doubly-ringed chemical structures called purines; the smaller nucleobases, cytosine and thymine (and uracil), are members of a class of singly-ringed chemical structures called pyrimidines.

Purines are only complementary with pyrimidines: pyrimidine-pyrimidine pairings are energetically unfavourable because the molecules are too far apart for hydrogen bonding to be established; purine-purine pairings are energetically unfavourable because the molecules are too close, leading to overlap repulsion. The only other possible pairings are GT and AC; these pairings are mismatches because the pattern of hydrogen donors and acceptors do not correspond. The GU pairing, with two hydrogen bonds, does occur fairly often in RNA.

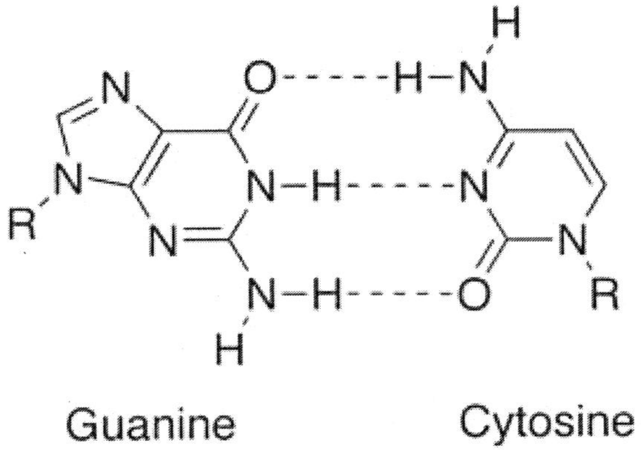

Figure: *A GC base pair demonstrating three intermolecular hydrogen bonds*

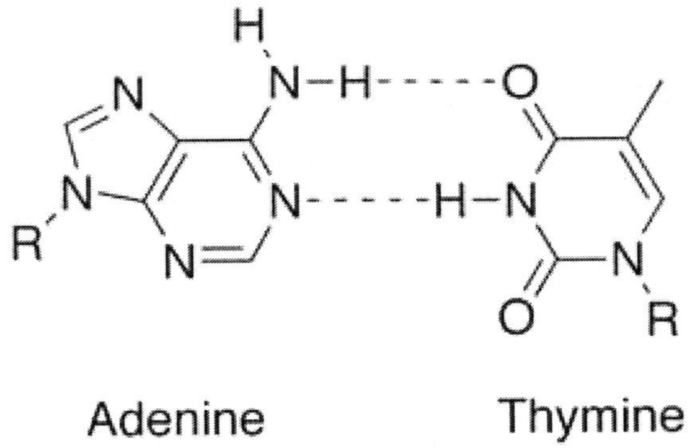

Figure: *An AT base pair demonstrating two intermolecular hydrogen bonds*

Paired DNA and RNA molecules are comparatively stable at room temperature but the two nucleotide strands will separate above a melting point that is determined by the length of the molecules, the extent of mispairing (if any), and the GC content. Higher GC content results in higher melting temperatures; it is therefore unsurprising that the genomes of extremophile organisms such as *Thermus thermophilus* are particularly GC-rich.

Conversely, regions of a genome that need to separate frequently—for example, the promoter regions for often-transcribed genes—are comparatively GC-poor. GC content and melting temperature must also be taken into account when designing primers for PCR reactions.

Base Stacking

Base stacking interactions in DNA and RNA are due to dispersion attraction, short-range exchange repulsion, and electrostatic interactions which also contribute to stability. Again GC stacking interactions with adjacent bases tend to be more favourable. (Note, though, that a GC stacking interaction with the next base pair is geometrically different from a CG interaction.) Base stacking effects are especially important in the secondary structure and tertiary structure of RNA; for example, RNA stem-loop structures are stabilized by base stacking in the loop region.

Base Analogs and Intercalators

Chemical analogs of nucleotides can take the place of proper nucleotides and establish non-canonical base-pairing, leading to errors (mostly point mutations) in DNA replication and DNA transcription. One common mutagenic base analog is 5-bromouracil, which resembles thymine but can base-pair to guanine in its enol form. Other chemicals, known as DNA intercalators, fit into the gap between adjacent bases on a single strand and induce frameshift mutations by "masquerading" as a base, causing the DNA replication machinery to skip or insert additional nucleotides at the intercalated site. Most intercalators are large polyaromatic compounds and are known or suspected carcinogens. Examples include ethidium bromide and acridine.

Examples

The following DNA sequences illustrate pair double-stranded patterns. By convention, the top strand is written from the 5' end to the 3' end; thus the bottom strand is written 3' to 5'.

A base-paired DNA sequence:

ATCGATTGAGCTCTAGCG

TAGCTAACTCGAGATCGC

The corresponding RNA sequence, in which uracil is substituted for thymine where uracil takes its place in the RNA strand:

AUCGAUUGAGCUCUAGCG

UAGCUAACUCGAGAUCGC

Length Measurements

The following abbreviations are commonly used to describe the length of a D/RNA molecule:

- bp = base pair(s)—one bp corresponds to circa 3.4 Å of length along the strand
- kb (= kbp) = kilo base pairs = 1,000 bp
- Mb = mega base pairs = 1,000,000 bp
- Gb = giga base pairs = 1,000,000,000 bp.

In case of single stranded DNA/RNA units of nucleotides are used, abbreviated nt (or knt, Mnt, Gnt), as they are not paired. For distinction between units of computer storage and bases kbp, Mbp, Gbp etc. may be used for basepairs. The length of 16S rDNA for bacteria is 1542 base-pair in length.

The Centimorgan is also often used to imply distance along a chromosome, but the number of base-pairs it corresponds to varies widely. In the Human genome, it is about 1 million base pairs

Sense and Antisense

A DNA sequence is called "sense" if its sequence is the same as that of a messenger RNA copy that is translated into protein. The sequence on the opposite strand is called the "antisense" sequence. Both sense and antisense sequences can exist on different parts of the same strand of DNA (i.e. both strands contain both sense and antisense sequences). In both prokaryotes and eukaryotes, antisense RNA sequences are produced, but the functions of these RNAs are not entirely clear. One proposal is that antisense RNAs are involved in regulating gene expression through RNA-RNA base pairing.

A few DNA sequences in prokaryotes and eukaryotes, and more in plasmids and viruses, blur the distinction between sense and antisense strands by having overlapping genes. In these cases, some DNA sequences do double duty, encoding one protein when read along one strand, and a second protein when read in the opposite direction along the other strand. In bacteria, this overlap may be involved in the regulation of gene transcription, while in viruses, overlapping genes increase the amount of information that can be encoded within the small viral genome.

Supercoiling

DNA can be twisted like a rope in a process called DNA supercoiling. With DNA in its "relaxed" state, a strand usually circles the axis of the double helix once every 10.4 base pairs, but if the DNA is twisted the strands become more tightly or more loosely wound. If the DNA is twisted in the direction of the helix, this is positive supercoiling, and the bases are held more tightly together. If they are twisted in the opposite direction, this is negative supercoiling, and the bases come apart more easily.

In nature, most DNA has slight negative supercoiling that is introduced by enzymes called topoisomerases. These enzymes are also needed to relieve the twisting stresses introduced into DNA strands during processes such as transcription and DNA replication.

Alternate DNA Structures

DNA exists in many possible conformations that include A-DNA, B-DNA, and Z-DNA forms, although, only B-DNA and Z-DNA have been directly observed in functional organisms. The conformation that DNA adopts depends on the hydration level, DNA sequence, the amount and direction of supercoiling, chemical modifications of the bases, the type and concentration of metal ions, as well as the presence of polyamines in solution.

The first published reports of A-DNA X-ray diffraction patterns—and also B-DNA used analyses based on Patterson transforms that provided only a limited amount of structural information for oriented fibers of DNA.

An alternate analysis was then proposed by Wilkins *et al.*, in 1953, for the *in vivo* B-DNA X-ray diffraction/scattering patterns of highly hydrated DNA fibers in terms of squares of Bessel functions. In the same journal, James D. Watson and Francis Crick presented their molecular modeling analysis of the DNA X-ray diffraction patterns to suggest that the structure was a double-helix.

Although the 'B-DNA form' is most common under the conditions found in cells, it is not a well-defined conformation but a family of related DNA conformations that occur at the high hydration levels present in living cells. Their corresponding X-ray diffraction and scattering patterns are characteristic of molecular paracrystals with a significant degree of disorder.

Compared to B-DNA, the A-DNA form is a wider right-handed spiral, with a shallow, wide minor groove and a narrower, deeper major groove. The A form occurs under non-physiological conditions in partially dehydrated samples of DNA, while in the cell it may be produced in hybrid pairings of DNA and RNA strands, as well as in enzyme-DNA complexes. Segments of DNA where the bases have been chemically modified by methylation may undergo a larger change in conformation and adopt the Z form. Here, the strands turn about the helical axis in a left-handed spiral, the opposite of the more common B form. These unusual structures can be recognized by specific Z-DNA binding proteins and may be involved in the regulation of transcription.

Molecular Models of DNA

Molecular models of DNA structures are representations of the molecular geometry and topology of Deoxyribonucleic acid (DNA) molecules using one of several means, with the aim of simplifying and presenting the essential, physical and chemical, properties of DNA molecular structures either *in vivo* or *in vitro*.

These representations include closely packed spheres (CPK models) made of plastic, metal wires for 'skeletal models', graphic computations and animations by computers, artistic rendering. Computer molecular models also allow animations and molecular dynamics simulations that are very important for understanding how DNA functions *in vivo*.

The more advanced, computer-based molecular models of DNA involve molecular dynamics simulations as well as quantum mechanical computations of vibro-rotations, delocalized molecular orbitals (MOs), electric dipole moments, hydrogen-bonding, and so on. DNA molecular dynamics modeling involves simulations of DNA molecular geometry and topology changes with time as a result of both intra- and inter-molecular interactions of DNA.

Whereas molecular models of Deoxyribonucleic acid (DNA) molecules such as closely packed spheres (CPK models) made of plastic or metal wires for 'skeletal models' are useful representations of static DNA structures, their usefulness is very limited for representing complex DNA dynamics. Computer molecular modeling allows both animations and molecular dynamics simulations that are very important for understanding how DNA functions *in vivo*.

From the very early stages of structural studies of DNA by X-ray diffraction and biochemical means, molecular models such as the Watson-Crick double-helix model were successfully employed to solve the 'puzzle' of DNA structure, and also find how the latter relates to its key functions in living cells. The first high quality X-ray diffraction patterns of A-DNA were reported by Rosalind Franklin and Raymond Gosling in 1953. The first calculations of the Fourier transform of an atomic helix were reported one year earlier by Cochran, Crick and Vand, and were followed in 1953 by the computation of the Fourier transform of a coiled-coil by Crick.

Structural information is generated from X-ray diffraction studies of oriented DNA fibers with the help of molecular models of DNA that are combined with crystallographic and mathematical analysis of the X-ray patterns. The first reports of a double-helix molecular model of B-DNA structure were made by Watson and Crick in 1953.

Last-but-not-least, Maurice F. Wilkins, A. Stokes and H.R. Wilson, reported the first X-ray patterns of *in vivo* B-DNA in partially oriented salmon sperm heads. The development of the first correct double-helix molecular model of DNA by Crick and Watson may not have been possible without the biochemical evidence for the nucleotide base-pairing ([A—T]; [C—G]), or Chargaff's rules.Although such initial studies of DNA structures with the help of molecular models were essentially static, their consequences for explaining the *in vivo* functions of DNA were significant in the areas of protein biosynthesis and the quasi-universality of the genetic code.

Epigenetic transformation studies of DNA *in vivo* were however much slower to develop in spite of their importance for embryology, morphogenesis and cancer research. Such chemical dynamics and biochemical reactions of DNA are much more complex than the molecular dynamics of DNA physical interactions with water, ions and proteins/enzymes in living cells.

Importance

An old standing dynamic problem is how DNA "self-replication" takes place in living cells that should involve transient uncoiling of supercoiled DNA fibers. Although DNA consists of relatively rigid, very large elongated biopolymer molecules called "fibers" or chains (that are made of repeating nucleotide units of four basic types, attached to deoxyribose and phosphate groups), its molecular structure *in vivo* undergoes dynamic configuration changes that involve dynamically attached water molecules and ions.

Supercoiling, packing with histones in chromosome structures, and other such supramolecular aspects also involve *in vivo* DNA topology which is even more complex than DNA molecular geometry, thus turning molecular modeling of DNA into an especially challenging problem for both molecular biologists and biotechnologists.

Like other large molecules and biopolymers, DNA often exists in multiple stable geometries (that is, it exhibits conformational isomerism) and configurational, quantum states which are close to each other in energy on the potential energy surface of the DNA molecule. Such varying molecular geometries can also be computed, at least in principle, by employing *ab initio* quantum chemistry methods that can attain high accuracy for small molecules, although claims that acceptable accuracy can be also achieved for polynuclelotides, as well as DNA conformations, were recently made on the basis of VCD spectral data.

Such quantum geometries define an important class of *ab initio* molecular models of DNA whose exploration has barely started especially in connection with results obtained by VCD in solutions. More detailed comparisons with such *ab initio* quantum computations are in principle obtainable through 2D-FT NMR spectroscopy and relaxation studies of polynucleotide solutions or specifically labelled DNA, as for example with deuterium labels. In an interesting twist of roles, the DNA molecule itself was proposed to be utilized for quantum computing. Both DNA nanostructures as well as DNA 'computing' biochips have been built.

Fundamental Concepts

The chemical structure of DNA is insufficient to understand the complexity of the 3D structures of DNA. On the other hand, animated molecular models allow one to visually explore the three-dimensional (3D) structure of DNA. The DNA model shown (far right) is a space-filling, or CPK, model of the DNA double-helix. Animated molecular models, such as the wire, or skeletal, type shown at the top of this article, allow one to visually explore the three-dimensional (3D) structure of DNA. Another type of DNA model is the space-filling, or CPK, model.

The hydrogen bonding dynamics and proton exchange is very different by many orders of magnitude between the two systems of fully hydrated DNA and water molecules in ice. Thus, the DNA dynamics is complex, involving nanosecond and several tens of picosecond time scales, whereas that of liquid ice is on the picosecond time scale, and that of proton exchange in ice is on the millisecond time scale.

The proton exchange rates in DNA and attached proteins may vary from picosecond to nanosecond, minutes or years, depending on the exact locations of the exchanged protons in the large biopolymers. A simple harmonic oscillator 'vibration' is only an oversimplified dynamic representation of the longitudinal vibrations of the DNA intertwined helices which were found to be anharmonic rather than harmonic as often assumed in quantum dynamic simulations of DNA.

DNA Structure

The structure of DNA shows a variety of forms, both double-stranded and single-stranded. The mechanical properties of DNA, which are directly related to its structure, are a significant problem for cells. Every process which binds or reads DNA is able to use or modify the mechanical properties of DNA for purposes of recognition, packaging and modification.

The extreme length (a chromosome may contain a 10 cm long DNA strand), relative rigidity and helical structure of DNA has led to the evolution of histones and of enzymes such as topoisomerases and helicases to manage a cell's DNA. The properties of DNA are closely related to its molecular structure and sequence, particularly the weakness of the hydrogen bonds and electronic interactions that hold strands of DNA together compared to the strength of the bonds within each strand.

Experimental techniques which can directly measure the mechanical properties of DNA are relatively new, and high-resolution visualization in solution is often difficult. Nevertheless, scientists have uncovered large amount of data on the mechanical properties of this polymer, and the implications of DNA's mechanical properties on cellular processes is a topic of active current research.

The DNA found in many cells can be macroscopic in length- a few centimetres long for each human chromosome. Consequently, cells must compact or "package" DNA to carry it within them. In eukaryotes this is carried by spool-like proteins known as histones, around which DNA winds. It is the further compaction of this DNA-protein complex which produces the well known mitotic eukaryotic chromosomes.

Paracrystalline Lattice Models of B-DNA Structures

A paracrystalline lattice, or paracrystal, is a molecular or atomic lattice with significant amounts (e.g., larger than a few percent) of partial disordering of molecular arranegements. Limiting cases of the paracrystal model are nanostructures, such as glasses, liquids, etc., that may possess only local ordering and no global order. Liquid crystals also have paracrystalline rather than crystalline structures. Highly hydrated B-DNA occurs naturally in living cells in such a paracrystalline state, which is a dynamic one in spite of the relatively rigid DNA double-helix stabilized by parallel hydrogen bonds between the nucleotide base-pairs in the two complementary, helical DNA chains. For simplicity most DNA molecular models omit both water and ions dynamically bound to B-DNA, and are thus less useful for understanding the dynamic behaviours of B-DNA *in vivo*. The physical and mathematical analysis of X-ray and spectroscopic data for paracrystalline B-DNA is therefore much more complicated than that of crystalline, A-DNA X-ray diffraction patterns. The paracrystal model is also important for DNA technological applications such as DNA nanotechnology. Novel techniques that combine X-ray diffraction of DNA with X-ray microscopy in hydrated living cells are now also being developed.

Genomic and Biotechnology Applications of DNA Molecular Modeling

There are various uses of DNA molecular modeling in Genomics and Biotechnology research applications, from DNA repair to PCR and DNA nanostructures. Two-dimensional DNA junction arrays have been visualized by Atomic force microscopy. DNA molecular modeling has various uses in genomics and biotechnology, with research applications ranging from DNA repair to PCR and DNA nanostructures. These include computer molecular models of molecules as varied as RNA polymerase, an E. coli, bacterial DNA primase template suggesting very complex dynamics at the interfaces between the enzymes and the DNA template, and molecular models of the mutagenic, chemical interaction of potent carcinogen molecules with DNA. These are all represented in the gallery below.

Technological application include a DNA biochip and DNA nanostructures designed for DNA computing and other dynamic applications of DNA nanotechnology. The image at right is of self-assembled DNA nanostructures. The DNA "tile" structure in this image consists of four branched junctions oriented at 90° angles. Each tile consists of nine DNA oligonucleotides as shown; such tiles serve as the primary "building block" for the assembly of the DNA nanogrids shown in the AFM micrograph.

Quadruplex Structures

At the ends of the linear chromosomes are specialized regions of DNA called telomeres. The main function of these regions is to allow the cell to replicate chromosome ends using the enzyme telomerase, as the enzymes that normally replicate DNA cannot copy the extreme 32 ends of chromosomes. These specialized chromosome caps also help protect the DNA ends, and stop the DNA repair systems in the cell from treating them as damage to be corrected. In human cells, telomeres are usually lengths of single-stranded DNA containing several thousand repeats of a simple TTAGGG sequence. These guanine-rich sequences may stabilize chromosome ends by forming structures of stacked sets of four-base units, rather than the usual base pairs found in other DNA molecules.

Here, four guanine bases form a flat plate and these flat four-base units then stack on top of each other, to form a stable *G-quadruplex* structure. These structures are stabilized by hydrogen bonding between the edges of the bases and chelation of a metal ion in the centre of each four-base unit. Other structures can also be formed, with the central set of four bases coming from either a single strand folded around the bases, or several different parallel strands, each contributing one base to the central structure.

In addition to these stacked structures, telomeres also form large loop structures called telomere loops, or T-loops. Here, the single-stranded DNA curls around in a long circle stabilized by telomere-binding proteins. At the very end of the T-loop, the single-stranded telomere DNA is held onto a region of double-stranded DNA by the telomere strand disrupting the double-helical DNA and base pairing to one of the two strands. This triple-stranded structure is called a displacement loop or D-loop.

G-quadruplex

Nucleic acid sequences which are rich in guanine are capable of forming four-stranded structures called G-quadruplexes (also known as G-tetrads or G_4-DNA). These consist of a square arrangement of guanines (a tetrad), stabilized by Hoogsteen hydrogen bonding. They are further stabilized by the existence of a monovalent cation (especially potassium) in the centre of the tetrads. They can be formed of DNA, RNA, LNA and PNA, and may be intramolecular, bimolecular or tetramolecular. Depending on the direction of the strands or parts of a strand that form the tetrads, structures may be described as parallel or antiparallel.

Telomeric Quadruplexes

Telomeric repeats in a variety of organisms have been shown to form these structures *in vitro*, and they have also been shown to form *in vivo* in some cases. The human telomeric repeat (which is the same for all vertebrates) consists of many repeats of the sequence d(GGTTAG), and the quadruplexes formed by this structure have been well studied by NMR and X-ray crystal structure determination. The formation of these quadruplexes in telomeres has been shown to decrease the activity of the enzyme telomerase, which is responsible for maintaining length of telomeres and is involved in around 85% of all cancers. This is an active target of drug discovery.

Non-telomeric Quadruplexes

Recently, there has been increasing interest in quadruplexes in locations other than at the telomere. This was given a large boost by the work by Simonsson and Hurley on the proto-oncogene c-myc, which was shown to form a quadruplex in a nuclease hypersensitive region critical for gene activity. Since then, many other genes have been shown to have G-quadruplexes in their promoter regions, including the chicken β-globin gene, human ubiquitin-ligase RFP2 and the proto-oncogenes c-kit, bcl-2, VEGF, H-ras and N-ras. This list is ever-increasing.

Genome-wide surveys based on a quadruplex folding rule have been performed, which have identified 376,000 Putative Quadruplex Sequences (PQS) in the human genome, although not all of these probably form *in vivo*. A similar study has identified putative G-quadruplexes in prokaryotes.There are several possible models for how quadruplexes could control gene activity, either by upregulation or downregulation. G-quadruplex formation in or near a promoter blocking transcription of the gene, and hence de-activating it. In another model, quadruplex formed at the non-coding DNA strand helps to maintain an open conformation of the coding DNA strand and enhance an expression of the respective gene.

Ligands which Bind Quadruplexes

One way of inducing or stabilizing G-quadruplex formation, is to introduce a molecule which can bind to the G-quadruplex structure, and a number of ligands, both small molecules and proteins, have been developed which can do so. This has become an increasingly large field of research. A number of naturally occurring proteins have been identified which selectively bind to G-quadruplexes. These include the helicases implicated in Bloom's and Werner's syndromes and the Saccharomyces cerevisiae protein RAP1. An artificially derived three zinc finger protein called Gq1, which is specific for G-quadruplexes has also been developed, as have specific antibodies. Cationic porphyrins have been shown to bind intercalatively with G-quadruplexes, as well as the molecule telomestatin.

Quadruplex Prediction Techniques

Identifying and predicting sequences which have the capacity to form quadruplexes is an important tool in further understanding of their role. A rule for predicting the formation has been proposed, where sequences are predicted to fold based on the pattern $d(G_{3+}N_{1-7}G_{3+}N_{1-7}G_{3+}N_{1-7}G_{3+})$, where N is any base (including guanine). This rule has been widely used in on-line algorithms.

Branched DNA

In DNA fraying occurs when non-complementary regions exist at the end of an otherwise complementary double-strand of DNA. However, branched DNA can occur if a third strand of DNA is introduced and contains adjoining regions able to hybridize with the frayed regions of the pre-existing double-strand. Although the simplest example of branched DNA involves only three strands of DNA, complexes involving additional strands and multiple branches are also possible.

Base Modifications

The expression of genes is influenced by how the DNA is packaged in chromosomes, in a structure called chromatin. Base modifications can be involved in packaging, with regions that have low or no gene expression usually containing high levels of methylation of cytosine bases. For example, cytosine methylation, produces 5-methylcytosine, which is important for X-chromosome inactivation.

The average level of methylation varies between organisms- the worm *Caenorhabditis elegans* lacks cytosine methylation, while vertebrates have higher levels, with up to 1% of their DNA containing 5-methylcytosine. Despite the importance of 5-methylcytosine, it can deaminate to leave a thymine base, so methylated cytosines are particularly prone to mutations. Other base modifications include adenine methylation in bacteria, the presence of 5-hydroxymethylcytosine in the brain, and the glycosylation of uracil to produce the "J-base" in.astids.

Damage

DNA can be damaged by many sorts of mutagens, which change the DNA sequence. Mutagens include oxidizing agents, alkylating agents and also high-energy electromagnetic radiation such as ultraviolet light and X-rays. The type of DNA damage produced depends on the type of mutagen. For example, UV light can damage DNA by producing thymine dimers, which are cross-links between pyrimidine bases. On the other hand, oxidants such as free radicals or hydrogen peroxide produce multiple forms of damage, including base modifications, particularly of guanosine, and double-strand breaks. A typical human cell contains about 150,000 bases that have suffered oxidative damage. Of these oxidative lesions, the most dangerous are double-strand breaks, as these are difficult to repair and can produce point mutations, insertions and deletions from the DNA sequence, as well as chromosomal translocations.

Many mutagens fit into the space between two adjacent base pairs, this is called *intercalation*. Most intercalators are aromatic and planar molecules; examples include ethidium bromide, daunomycin, and doxorubicin. In order for an intercalator to fit between base pairs, the bases must separate, distorting the DNA strands by unwinding of the double helix.

This inhibits both transcription and DNA replication, causing toxicity and mutations. As a result, DNA intercalators are often carcinogens, and benzo[*a*]pyrene diol epoxide, acridines, aflatoxin and ethidium bromide are well-known examples. Nevertheless, due to their ability to inhibit DNA transcription and replication, other similar toxins are also used in chemotherapy to inhibit rapidly growing cancer cells.

Mutation

Mutations are changes in a genomic sequence: the DNA sequence of a cell's genome or the DNA or RNA sequence of a virus. Mutations are caused by radiation, viruses, transposons and mutagenic chemicals, as well as errors that occur during meiosis or DNA replication. They can also be induced by the organism itself, by cellular processes such as hypermutation.

Mutation can result in several different types of change in DNA sequences; these can either have no effect, alter the product of a gene, or prevent the gene from functioning properly or completely. Studies in the fly *Drosophila melanogaster* suggest that if a mutation changes a protein produced by a gene, this will probably be harmful, with about 70 percent of these mutations having damaging effects, and the remainder being either neutral or weakly beneficial. Due to the damaging effects that mutations can have on genes, organisms have evolved mechanisms such as DNA repair to remove mutations.

Therefore, the optimal mutation rate for a species is a trade-off between costs of a high mutation rate, such as deleterious mutations, and the metabolic costs of maintaining systems to reduce the mutation rate, such as DNA repair enzymes. Viruses that use RNA as their genetic material have rapid mutation rates, which can be an advantage since these viruses will evolve constantly and rapidly, and thus evade the defensive responses of e.g. the human immune system.

Description

Mutations can involve large sections of DNA becoming duplicated, usually through genetic recombination. These duplications are a major source of raw material for evolving new genes, with tens to hundreds of genes duplicated in animal genomes every million years. Most genes belong to larger families of genes of shared ancestry. Novel genes are produced by several methods, commonly through the duplication and mutation of an ancestral gene, or by recombining parts of different genes to form new combinations with new functions.

Here, domains act as modules, each with a particular and independent function, that can be mixed together to produce genes encoding new proteins with novel properties. For example, the human eye uses four genes to make structures that sense light: three for colour vision and one for night vision; all four arose from a single ancestral gene. Another advantage of duplicating a gene (or even an entire genome) is that this increases redundancy; this allows one gene in the pair to acquire a new function while the other copy performs the original function. Other types of mutation occasionally create new genes from previously noncoding DNA.

Changes in chromosome number may involve even larger mutations, where segments of the DNA within chromosomes break and then rearrange. For example, two chromosomes in the *Homo* genus fused to produce human chromosome 2; this fusion did not occur in the lineage of the other apes, and they retain these separate chromosomes. In evolution, the most important role of such chromosomal rearrangements may be to accelerate the divergence of a population into new species by making populations less likely to interbreed, and thereby preserving genetic differences between these populations.

Sequences of DNA that can move about the genome, such as transposons, make up a major fraction of the genetic material of plants and animals, and may have been important in the evolution of genomes. For example, more than a million copies of the Alu sequence are present in the human genome, and these sequences have now been recruited to perform functions such as regulating gene expression. Another effect of these mobile DNA sequences is that when they move within a genome, they can mutate or delete existing genes and thereby produce genetic diversity.

In multicellular organisms with dedicated reproductive cells, mutations can be subdivided into germ line mutations, which can be passed on to descendants through their reproductive cells, and somatic mutations (also called acquired mutations), which involve cells outside the dedicated reproductive group and which are not usually transmitted to descendants. If the organism can reproduce asexually through mechanisms such as cuttings or budding the distinction can become blurred. For example, plants can sometimes transmit somatic mutations to their descendants asexually or sexually where flower buds develop in somatically mutated parts of plants. A new mutation that was not inherited from either parent is called a *de novo* mutation. The source of the mutation is unrelated to the consequence, although the consequences are related to which cells were mutated.

Nonlethal mutations accumulate within the gene pool and increase the amount of genetic variation. The abundance of some genetic changes within the gene pool can be reduced by natural selection, while other "more favourable" mutations may accumulate and result in adaptive evolutionary changes. For example, a butterfly may produce offspring with new mutations. The majority of these mutations will have no effect; but one might change the colour of one of the butterfly's offspring, making it harder (or easier) for predators to see. If this colour change is advantageous, the chance of this butterfly surviving and producing its own offspring are a little better, and over time the number of butterflies with this mutation may form a larger percentage of the population.

Neutral mutations are defined as mutations whose effects do not influence the fitness of an individual. These can accumulate over time due to genetic drift. It is believed that the overwhelming majority of mutations have no significant effect on an organism's fitness. Also, DNA repair mechanisms are able to mend most changes before they become permanent mutations, and many organisms have mechanisms for eliminating otherwise permanently mutated somatic cells.

Mutation is generally accepted by biologists as the mechanism by which natural selection acts, generating advantageous new traits that survive and multiply in offspring as well as disadvantageous traits, in less fit offspring, that tend to die out.

Causes

Two classes of mutations are spontaneous mutations (molecular decay) and induced mutations caused by mutagens.

Spontaneous mutations on the molecular level can be caused by:

- Tautomerism – A base is changed by the repositioning of a hydrogen atom, altering the hydrogen bonding pattern of that base resulting in incorrect base pairing during replication.
- Depurination – Loss of a purine base (A or G) to form an apurinic site (AP site).
- Deamination – Hydrolysis changes a normal base to an atypical base containing a keto group in place of the original amine group. Examples include C → U and A → HX (hypoxanthine), which can be corrected by DNA repair mechanisms; and 5MeC (5-methylcytosine) → T, which is less likely to be detected as a mutation because thymine is a normal DNA base.
- Slipped strand mispairing-Denaturation of the new strand from the template during replication, followed by renaturation in a different spot ("slipping"). This can lead to insertions or deletions.

Induced mutations on the molecular level can be caused by:

- Chemicals:
  - Hydroxylamine NH_2OH
  - Base analogs (e.g. BrdU)
  - Alkylating agents (e.g. *N*-ethyl-*N*-nitrosourea) These agents can mutate both replicating and non-replicating DNA. In contrast, a base analog can only mutate the DNA when the analog is incorporated in replicating the DNA. Each of these classes of chemical mutagens has certain effects that then lead to transitions, transversions, or deletions.
  - Agents that form DNA adducts (e.g. ochratoxin A metabolites)
  - DNA intercalating agents (e.g. ethidium bromide)
  - DNA crosslinkers
  - Oxidative damage
  - Nitrous acid converts amine groups on A and C to diazo groups, altering their hydrogen bonding patterns which leads to incorrect base pairing during replication.

- Radiation:
  - o Ultraviolet radiation (nonionizing radiation). Two nucleotide bases in DNA – cytosine and thymine – are most vulnerable to radiation that can change their properties. UV light can induce adjacent pyrimidine bases in a DNA strand to become covalently joined as a pyrimidine dimer. UV radiation, particularly longer-wave UVA, can also cause oxidative damage to DNA.
  - o Ionizing radiation
- Viral infections.

DNA has so-called hotspots, where mutations occur up to 100 times more frequently than the normal mutation rate. A hotspot can be at an unusual base, e.g., 5-methylcytosine.

Mutation rates also vary across species. Evolutionary biologists have theorized that higher mutation rates are beneficial in some situations, because they allow organisms to evolve and therefore adapt more quickly to their environments.

For example, repeated exposure of bacteria to antibiotics, and selection of resistant mutants, can result in the selection of bacteria that have a much higher mutation rate than the original population (mutator strains).

Classification of Mutation Types

The sequence of a gene can be altered in a number of ways. Gene mutations have varying effects on health depending on where they occur and whether they alter the function of essential proteins. Mutations in the structure of genes can be classified as:

- Small-scale mutations, such as those affecting a small gene in one or a few nucleotides, including:
  - o Point mutations, often caused by chemicals or malfunction of DNA replication, exchange a single nucleotide for another. These changes are classified as transitions or transversions. Most common is the transition that exchanges a purine for a purine (A "! G) or a pyrimidine for a pyrimidine, (C $\rightarrow$ T). A transition can be caused by nitrous acid, base mis-pairing, or mutagenic base analogs such as 5-bromo-2-deoxyuridine (BrdU). Less common is a transversion, which exchanges a purine for a pyrimidine or a pyrimidine for a purine (C/T $\rightarrow$ A/G). An example of a transversion is adenine

(A) being converted into a cytosine (C). A point mutation can be reversed by another point mutation, in which the nucleotide is changed back to its original state (true reversion) or by second-site reversion (a complementary mutation elsewhere that results in regained gene functionality). Point mutations that occur within the protein coding region of a gene may be classified into three kinds, depending upon what the erroneous codon codes for:

  - — Silent mutations: which code for the same amino acid.
  - — Missense mutations: which code for a different amino acid.
  - — Nonsense mutations: which code for a stop and can truncate the protein.

- o Insertions add one or more extra nucleotides into the DNA. They are usually caused by transposable elements, or errors during replication of repeating elements (e.g. AT repeats). Insertions in the coding region of a gene may alter splicing of the mRNA (splice site mutation), or cause a shift in the reading frame (frameshift), both of which can significantly alter the gene product. Insertions can be reverted by excision of the transposable element.
- o Deletions remove one or more nucleotides from the DNA. Like insertions, these mutations can alter the reading frame of the gene. They are generally irreversible: though exactly the same sequence might theoretically be restored by an insertion, transposable elements able to revert a very short deletion (say 1–2 bases) in *any* location are either highly unlikely to exist or do not exist at all. Note that a deletion is not the exact opposite of an insertion: the former is quite random while the latter consists of a specific sequence inserting at locations that are not entirely random or even quite narrowly defined.

- Large-scale mutations in chromosomal structure, including:
  - o Amplifications (or gene duplications) leading to multiple copies of all chromosomal regions, increasing the dosage of the genes located within them.
  - o Deletions of large chromosomal regions, leading to loss of the genes within those regions.

- o Mutations whose effect is to juxtapose previously separate pieces of DNA, potentially bringing together separate genes to form functionally distinct fusion genes (e.g. bcr-abl). These include:
  - — Chromosomal translocations: interchange of genetic parts from nonhomologous chromosomes.
  - — Interstitial deletions: an intra-chromosomal deletion that removes a segment of DNA from a single chromosome, thereby apposing previously distant genes. For example, cells isolated from a human astrocytoma, a type of brain tumour, were found to have a chromosomal deletion removing sequences between the "fused in glioblastoma" gene and the receptor tyrosine kinase "ros", producing a fusion protein (FIG-ROS). The abnormal FIG-ROS fusion protein has constitutively active kinase activity that causes oncogenic transformation (a transformation from normal cells to cancer cells).
  - — Chromosomal inversions: reversing the orientation of a chromosomal segment.
- o Loss of heterozygosity: loss of one allele, either by a deletion or recombination event, in an organism that previously had two different alleles.

By Effect on Function

- Loss-of-function mutations are the result of gene product having less or no function. When the allele has a complete loss of function (null allele) it is often called an amorphic mutation. Phenotypes associated with such mutations are most often recessive. Exceptions are when the organism is haploid, or when the reduced dosage of a normal gene product is not enough for a normal phenotype (this is called aploinsufficiency).
- Gain-of-function mutations change the gene product such that it gains a new and abnormal function. These mutations usually have dominant phenotypes. Often called a neomorphic mutation.
- Dominant negative mutations (also called antimorphic mutations) have an altered gene product that acts antagonistically to the wild-type allele. These mutations usually result in an altered molecular function (often inactive) and are characterised by a dominant or semi-dominant phenotype. In

humans, Marfan syndrome is an example of a dominant negative mutation occurring in an autosomal dominant disease. In this condition, the defective glycoprotein product of the fibrillin gene (FBN1) antagonizes the product of the normal allele.

- Lethal mutations are mutations that lead to the death of the organisms which carry the mutations.
- A back mutation or reversion is a point mutation that restores the original sequence and hence the original phenotype.

By Effect on Fitness

In applied genetics it is usual to speak of mutations as either harmful or beneficial.

- A harmful mutation is a mutation that decreases the fitness of the organism.
- A beneficial mutation is a mutation that increases fitness of the organism, or which promotes traits that are desirable.

In theoretical population genetics, it is more usual to speak of such mutations as deleterious or advantageous. In the neutral theory of molecular evolution, genetic drift is the basis for most variation at the molecular level.

- A neutral mutation has no harmful or beneficial effect on the organism. Such mutations occur at a steady rate, forming the basis for the molecular clock.
- A deleterious mutation has a negative effect on the phenotype, and thus decreases the fitness of the organism.
- An advantageous mutation has a positive effect on the phenotype, and thus increases the fitness of the organism.
- A nearly neutral mutation is a mutation that may be slightly deleterious or advantageous, although most nearly neutral mutations are slightly deleterious.

In reality, viewing the fitness effects of mutations in these discrete categories is an oversimplification. Attempts have been made to infer the distribution of fitness effects using mutagenesis experiments or theoretical models applied to molecular sequence data. However, the current distribution is still uncertain, and some aspects of the distribution likely vary between species.

By Inheritance:

- inheritable generic in pro-generic tissue or cells on path to be changed to gametes.

- non inheritable somatic (e.g., carcinogenic mutation)
- non inheritable post mortem aDNA mutation in decaying remains.

By pattern of inheritance The human genome contains two copies of each gene – a paternal and a maternal allele.

- A heterozygous mutation is a mutation of only one allele.
- A homozygous mutation is an identical mutation of both the paternal and maternal alleles.
- Compound heterozygous mutations or a genetic compound comprises two different mutations in the paternal and maternal alleles.
- A wildtype or homozygous non-mutated organism is one in which neither allele is mutated. (Just not a mutation)

By Impact on Protein Sequence

- A frameshift mutation is a mutation caused by insertion or deletion of a number of nucleotides that is not evenly divisible by three from a DNA sequence. Due to the triplet nature of gene expression by codons, the insertion or deletion can disrupt the reading frame, or the grouping of the codons, resulting in a completely different translation from the original. The earlier in the sequence the deletion or insertion occurs, the more altered the protein produced is.
- A nonsense mutation is a point mutation in a sequence of DNA that results in a premature stop codon, or a *nonsense codon* in the transcribed mRNA, and possibly a truncated, and often nonfunctional protein product.
- Missense mutations or *nonsynonymous mutations* are types of point mutations where a single nucleotide is changed to cause substitution of a different amino acid. This in turn can render the resulting protein nonfunctional. Such mutations are responsible for diseases such as Epidermolysis bullosa, sickle-cell disease, and SOD1 mediated ALS.
- A neutral mutation is a mutation that occurs in an amino acid codon which results in the use of a different, but chemically similar, amino acid. The similarity between the two is enough that little or no change is often rendered in the protein. For example, a change from AAA to AGA will encode lysine, a chemically similar molecule to the intended arginine.

- Silent mutations are mutations that do not result in a change to the amino acid sequence of a protein. They may occur in a region that does not code for a protein, or they may occur within a codon in a manner that does not alter the final amino acid sequence. The phrase *silent mutation* is often used interchangeably with the phrase *synonymous mutation*; however, synonymous mutations are a subcategory of the former, occurring only within exons. The name silent could be a misnomer. For example, a silent mutation in the exon/intron border may lead to alternative splicing by changing the splice site, thereby leading to a changed protein.

Special Classes

- Conditional mutation is a mutation that has wild-type (or less severe) phenotype under certain "permissive" environmental conditions and a mutant phenotype under certain "restrictive" conditions. For example, a temperature-sensitive mutation can cause cell death at high temperature (restrictive condition), but might have no deleterious consequences at a lower temperature (permissive condition).

Nomenclature

Nomenclature of mutations specify the type of mutation and base or amino acid changes.

- Nucleotide substitution (e.g. 76A>T)- The number is the position of the nucleotide from the 5' end, the first letter represents the wild type nucleotide, and the second letter represents the nucleotide which replaced the wild type. In the given example, the adenine at the 76th position was replaced by a thymine.
  - If it becomes necessary to differentiate between mutations in genomic DNA, mitochondrial DNA, and RNA, a simple convention is used. For example, if the 100th base of a nucleotide sequence mutated from G to C, then it would be written as g.100G>C if the mutation occurred in genomic DNA, m.100G>C if the mutation occurred in mitochondrial DNA, or r.100g>c if the mutation occurred in RNA. Note that for mutations in RNA, the nucleotide code is written in lower case.
- Amino acid substitution (e.g. D111E) – The first letter is the one letter code of the wild type amino acid, the number is the position of the amino acid from the N terminus, and the second

letter is the one letter code of the amino acid present in the mutation. Nonsense mutations are represented with an X for the second amino acid (e.g. D111X).

- Amino acid deletion (e.g. ΔF508) – The Greek letter Δ (delta) indicates a deletion. The letter refers to the amino acid present in the wild type and the number is the position from the N terminus of the amino acid were it to be present as in the wild type.

Harmful Mutations

Changes in DNA caused by mutation can cause errors in protein sequence, creating partially or completely non-functional proteins. To function correctly, each cell depends on thousands of proteins to function in the right places at the right times. When a mutation alters a protein that plays a critical role in the body, a medical condition can result. A condition caused by mutations in one or more genes is called a genetic disorder. Some mutations alter a gene's DNA base sequence but do not change the function of the protein made by the gene. Studies of the fly *Drosophila melanogaster* suggest that if a mutation does change a protein, this will probably be harmful, with about 70 percent of these mutations having damaging effects, and the remainder being either neutral or weakly beneficial. However, studies in yeast have shown that only 7% of mutations that are not in genes are harmful.

If a mutation is present in a germ cell, it can give rise to offspring that carries the mutation in all of its cells. This is the case in hereditary diseases. On the other hand, a mutation may occur in a somatic cell of an organism. Such mutations will be present in all descendants of this cell within the same organism, and certain mutations can cause the cell to become malignant, and thus cause cancer.

Often, gene mutations that could cause a genetic disorder are repaired by the DNA repair system of the cell. Each cell has a number of pathways through which enzymes recognize and repair mistakes in DNA. Because DNA can be damaged or mutated in many ways, the process of DNA repair is an important way in which the body protects itself from disease.

Beneficial Mutations

Although most mutations that change protein sequences are neutral or harmful, some mutations have a positive effect on an organism. In this case, the mutation may enable the mutant organism to withstand particular environmental stresses better than wild-type organisms, or reproduce more quickly. In these cases a mutation will tend to become more common in a population through natural selection.

For example, a specific 32 base pair deletion in human CCR5 (CCR5-Δ32) confers HIV resistance to homozygotes and delays AIDS onset in heterozygotes. The CCR5 mutation is more common in those of European descent. One possible explanation of the etiology of the relatively high frequency of CCR5-Δ32 in the European population is that it conferred resistance to the bubonic plague in mid-14th century Europe. People with this mutation were more likely to survive infection; thus its frequency in the population increased. This theory could explain why this mutation is not found in southern Africa, where the bubonic plague never reached. A newer theory suggests that the selective pressure on the CCR5 Delta 32 mutation was caused by smallpox instead of the bubonic plague.

Another example, is Sickle cell disease which is a blood disorder in which the body produces an abnormal type of the oxygen-carrying substance hemoglobin in the red blood cells. One-third of all indigenous inhabitants of Sub-Saharan Africa carry the gene, because in areas where malaria is common, there is a survival value in carrying only a single sickle-cell gene (sickle cell trait). Those with only one of the two alleles of the sickle-cell disease are more resistant to malaria, since the infestation of the malaria plasmodium is halted by the sickling of the cells which it infests.

Prion Mutation

Prions are proteins and do not contain genetic material. However, prion replication has been shown to be subject to mutation and natural selection just like other forms of replication.

Biological Functions

DNA usually occurs as linear chromosomes in eukaryotes, and circular chromosomes in prokaryotes. The set of chromosomes in a cell makes up its genome; the human genome has approximately 3 billion base pairs of DNA arranged into 46 chromosomes. The information carried by DNA is held in the sequence of pieces of DNA called genes. Transmission of genetic information in genes is achieved via complementary base pairing. For example, in transcription, when a cell uses the information in a gene, the DNA sequence is copied into a complementary RNA sequence through the attraction between the DNA and the correct RNA nucleotides. Usually, this RNA copy is then used to make a matching protein sequence in a process called translation, which depends on the same interaction between RNA nucleotides. In alternative fashion, a cell may simply copy its genetic information in a process called DNA replication. The details of these functions are covered in other articles; here we focus on the interactions between DNA and other molecules that mediate the function of the genome.

Genes and Genomes

Genomic DNA is tightly and orderly packed in the process called DNA condensation to fit the small available volumes of the cell. In eukaryotes, DNA is located in the cell nucleus, as well as small amounts in mitochondria and chloroplasts. In prokaryotes, the DNA is held within an irregularly shaped body in the cytoplasm called the nucleoid. The genetic information in a genome is held within genes, and the complete set of this information in an organism is called its genotype. A gene is a unit of heredity and is a region of DNA that influences a particular characteristic in an organism. Genes contain an open reading frame that can be transcribed, as well as regulatory sequences such as promoters and enhancers, which control the transcription of the open reading frame.

In many species, only a small fraction of the total sequence of the genome encodes protein. For example, only about 1.5% of the human genome consists of protein-coding exons, with over 50% of human DNA consisting of non-coding repetitive sequences. The reasons for the presence of so much non-coding DNA in eukaryotic genomes and the extraordinary differences in genome size, or *C-value*, among species represent a long-standing puzzle known as the "C-value enigma". However, DNA sequences that do not code protein may still encode functional non-coding RNA molecules, which are involved in the regulation of gene expression. Some non-coding DNA sequences play structural roles in chromosomes. Telomeres and centromeres typically contain few genes, but are important for the function and stability of chromosomes. An abundant form of non-coding DNA in humans are pseudogenes, which are copies of genes that have been disabled by mutation. These sequences are usually just molecular fossils, although they can occasionally serve as raw genetic material for the creation of new genes through the process of gene duplication and divergence.

Cell Nucleus

The nucleus is a membrane enclosed organelle found in eukaryotic cells. It contains most of the cell's genetic material, organized as multiple long linear DNA molecules in complex with a large variety of proteins, such as histones, to form chromosomes. The genes within these chromosomes are the cell's nuclear genome. The function of the nucleus is to maintain the integrity of these genes and to control the activities of the cell by regulating gene expression — the nucleus is therefore the control centre of the cell.

The main structures making up the nucleus are the nuclear envelope, a double membrane that encloses the entire organelle and separates its contents from the cellular cytoplasm, and the nuclear lamina, a meshwork within the nucleus that adds mechanical support, much like the cytoskeleton supports the cell as a whole.

Because the nuclear membrane is impermeable to most molecules, nuclear pores are required to allow movement of molecules across the envelope. These pores cross both of the membranes, providing a channel that allows free movement of small molecules and ions. The movement of larger molecules such as proteins is carefully controlled, and requires active transport regulated by carrier proteins. Nuclear transport is crucial to cell function, as movement through the pores is required for both gene expression and chromosomal maintenance.

Although the interior of the nucleus does not contain any membrane-bound subcompartments, its contents are not uniform, and a number of *subnuclear bodies* exist, made up of unique proteins, RNA molecules, and particular parts of the chromosomes. The best known of these is the nucleolus, which is mainly involved in the assembly of ribosomes. After being produced in the nucleolus, ribosomes are exported to the cytoplasm where they translate mRNA.

Structures

The nucleus is the largest cellular organelle in animals. In mammalian cells, the average diameter of the nucleus is approximately 6 micrometers (μm), which occupies about 10% of the total cell volume. The viscous liquid within it is called nucleoplasm, and is similar in composition to the cytosol found outside the nucleus. It appears as a dense, roughly spherical organelle.

Nuclear Envelope and Pores

The nuclear envelope otherwise known as nuclear membrane consists of two cellular membranes, an inner and an outer membrane, arranged parallel to one another and separated by 10 to 50 manometers (nm). The nuclear envelope completely encloses the nucleus and separates the cell's genetic material from the surrounding cytoplasm, serving as a barrier to prevent macromolecules from diffusing freely between the nucleoplasm and the cytoplasm. The outer nuclear membrane is continuous with the membrane of the rough endoplasmic reticulum (RER), and is similarly studded with ribosomes. The space between the membranes is called the perinuclear space and is continuous with the RER lumen.

Nuclear pores, which provide aqueous channels through the envelope, are composed of multiple proteins, collectively referred to as nucleoporins. The pores are about 125 million daltons in molecular weight and consist of around 50 (in yeast) to 100 proteins (in vertebrates). The pores are 100 nm in total diameter; however, the gap through which molecules freely diffuse is only about 9 nm wide, due to the presence of regulatory systems within the centre of the pore. This size allows the free passage of small water-soluble molecules while preventing larger molecules, such as nucleic acids and larger proteins, from inappropriately entering or exiting the nucleus. These large molecules must be actively transported into the nucleus instead. The nucleus of a typical mammalian cell will have about 3000 to 4000 pores throughout its envelope, each of which contains a donut-shaped, eightfold-symmetric ring-shaped structure at a position where the inner and outer membranes fuse. Attached to the ring is a structure called the *nuclear basket* that extends into the nucleoplasm, and a series of filamentous extensions that reach into the cytoplasm. Both structures serve to mediate binding to nuclear transport proteins.

Most proteins, ribosomal subunits, and some RNAs are transported through the pore complexes in a process mediated by a family of transport factors known as karyopherins. Those karyopherins that mediate movement into the nucleus are also called importins, while those that mediate movement out of the nucleus are called exportins. Most karyopherins interact directly with their cargo, although some use adaptor proteins. Steroid hormones such as cortisol and aldosterone, as well as other small lipid-soluble molecules involved in intercellular signalling can diffuse through the cell membrane and into the cytoplasm, where they bind nuclear receptor proteins that are trafficked into the nucleus. There they serve as transcription factors when bound to their ligand; in the absence of ligand many such receptors function as histone deacetylases that repress gene expression.

Nuclear Lamina

In animal cells, two networks of intermediate filaments provide the nucleus with mechanical support: the nuclear lamina forms an organized meshwork on the internal face of the envelope, while less organized support is provided on the cytosolic face of the envelope. Both systems provide structural support for the nuclear envelope and anchoring sites for chromosomes and nuclear pores.

The nuclear lamina is mostly composed of lamin proteins. Like all proteins, lamins are synthesized in the cytoplasm and later transported into the nucleus interior, where they are assembled before being incorporated into the existing network of nuclear lamina. Lamins are also found inside the nucleoplasm where they form another regular structure, known as the *nucleoplasmic veil*, that is visible using fluorescence microscopy. The actual function of the veil is not clear, although it is excluded from the nucleolus and is present during interphase. The lamin structures that make up the veil bind chromatin and disrupting their structure inhibits transcription of protein-coding genes.

Like the components of other intermediate filaments, the lamin monomer contains an alpha-helical domain used by two monomers to coil around each other, forming a dimer structure called a coiled coil. Two of these dimer structures then join side by side, in an antiparallel arrangement, to form a tetramer called a *protofilament*. Eight of these protofilaments form a lateral arrangement that is twisted to form a ropelike *filament*. These filaments can be assembled or disassembled in a dynamic manner, meaning that changes in the length of the filament depend on the competing rates of filament addition and removal.

Mutations in lamin genes leading to defects in filament assembly are known as *laminopathies*. The most notable laminopathy is the family of diseases known as progeria, which causes the appearance of premature aging in its sufferers. The exact mechanism by which the associated biochemical changes give rise to the aged phenotype is not well understood.

Chromosomes

A chromosome is an organized structure of DNA and protein that is found in cells. It is a single piece of coiled DNA containing many genes, regulatory elements and other nucleotide sequences. Chromosomes also contain DNA-bound proteins, which serve to package the DNA and control its functions. Chromosomes vary widely between different organisms. The DNA molecule may be circular or linear, and can be composed of 10,000 to 1,000,000,000 nucleotides in a long chain. Typically eukaryotic cells (cells with nuclei) have large linear chromosomes and prokaryotic cells (cells without defined nuclei) have smaller circular chromosomes, although there are many exceptions to this rule. Furthermore, cells may contain more than one type of chromosome; for example, mitochondria in most eukaryotes and chloroplasts in plants have their own small chromosomes.

In eukaryotes, nuclear chromosomes are packaged by proteins into a condensed structure called chromatin. This allows the very long DNA molecules to fit into the cell nucleus. The structure of chromosomes and chromatin varies through the cell cycle. Chromosomes are the essential unit for cellular division and must be replicated, divided, and passed successfully to their daughter cells so as to ensure the genetic diversity and survival of their progeny. Chromosomes may exist as either duplicated or unduplicated—unduplicated chromosomes are single linear strands, whereas duplicated chromosomes (copied during synthesis phase) contain two copies joined by a centromere.

Compaction of the duplicated chromosomes during mitosis and meiosis results in the classic four-arm structure (pictured to the right). Chromosomal recombination plays a vital role in genetic diversity. If these structures are manipulated incorrectly, through processes known as chromosomal instability and translocation, the cell may undergo mitotic catastrophe and die, or it may unexpectedly evade apoptosis leading to the progression of cancer.

In practice "chromosome" is a rather loosely defined term. In prokaryotes and viruses, the term genophore is more appropriate when no chromatin is present. However, a large body of work uses the term chromosome regardless of chromatin content. In prokaryotes DNA is usually arranged as a circle, which is tightly coiled in on itself, sometimes accompanied by one or more smaller, circular DNA molecules called plasmids. These small circular genomes are also found in mitochondria and chloroplasts, reflecting their bacterial origins. The simplest genophores are found in viruses: these DNA or RNA molecules are short linear or circular genophores that often lack structural proteins.

Chromosomes in Eukaryotes

Eukaryotes (cells with nuclei such as those found in plants, yeast, and animals) possess multiple large linear chromosomes contained in the cell's nucleus. Each chromosome has one centromere, with one or two arms projecting from the centromere, although, under most circumstances, these arms are not visible as such. In addition, most eukaryotes have a small circular mitochondrial genome, and some eukaryotes may have additional small circular or linear cytoplasmic chromosomes. In the nuclear chromosomes of eukaryotes, the uncondensed DNA exists in a semi-ordered structure, where it is wrapped around histones (structural proteins), forming a composite material called chromatin.

Chromatin

Chromatin is the complex of DNA and protein found in the eukaryotic nucleus, which packages chromosomes. The structure of chromatin varies significantly between different stages of the cell cycle, according to the requirements of the DNA.

Interphase Chromatin

During interphase (the period of the cell cycle where the cell is not dividing), two types of chromatin can be distinguished:

- Euchromatin, which consists of DNA that is active, e.g., being expressed as protein.
- Heterochromatin, which consists of mostly inactive DNA. It seems to serve structural purposes during the chromosomal stages. Heterochromatin can be further distinguished into two types:
  - *Constitutive heterochromatin*, which is never expressed. It is located around the centromere and usually contains repetitive sequences.
  - *Facultative heterochromatin*, which is sometimes expressed.

Individual chromosomes cannot be distinguished at this stage – they appear in the nucleus as a homogeneous tangled mix of DNA and protein.

Metaphase Chromatin and Division

In the early stages of mitosis or meiosis (cell division), the chromatin strands become more and more condensed. They cease to function as accessible genetic material (transcription stops) and become a compact transportable form. This compact form makes the individual chromosomes visible, and they form the classic four arm structure, a pair of sister chromatids attached to each other at the centromere. The shorter arms are called *p arms* (from the French *petit*, small) and the longer arms are called *q arms* (*q* follows *p* in the Latin alphabet). This is the only natural context in which individual chromosomes are visible with an optical microscope.

During divisions, long microtubules attach to the centromere and the two opposite ends of the cell. The microtubules then pull the chromatids apart, so that each daughter cell inherits one set of chromatids. Once the cells have divided, the chromatids are uncoiled and can function again as chromatin. In spite of their appearance, chromosomes are structurally highly condensed, which enables these giant DNA structures to be contained within a cell nucleus.

The self-assembled microtubules form the spindle, which attaches to chromosomes at specialized structures called kinetochores, one of which is present on each sister chromatid. A special DNA base sequence in the region of the kinetochores provides, along with special proteins, longer-lasting attachment in this region.

Chromosomes in Prokaryotes

The prokaryotes – bacteria and archaea – typically have a single circular chromosome, but many variations do exist. Most bacteria have a single circular chromosome that can range in size from only 160,000 base pairs in the endosymbiotic bacterium *Candidatus Carsonella ruddii*, to 12,200,000 base pairs in the soil-dwelling bacterium *Sorangium cellulosum*. Spirochaetes of the genus *Borrelia* are a notable exception to this arrangement, with bacteria such as *Borrelia burgdorferi*, the cause of Lyme disease, containing a single linear chromosome.

Structure in Sequences

Prokaryotic chromosomes have less sequence-based structure than eukaryotes. Bacteria typically have a single point (the origin of replication) from which replication starts, whereas some archaea contain multiple replication origins. The genes in prokaryotes are often organized in operons, and do not usually contain introns, unlike eukaryotes.

DNA Packaging

Prokaryotes do not possess nuclei. Instead, their DNA is organized into a structure called the nucleoid. The nucleoid is a distinct structure and occupies a defined region of the bacterial cell. This structure is, however, dynamic and is maintained and remodelled by the actions of a range of histone-like proteins, which associate with the bacterial chromosome. In archaea, the DNA in chromosomes is even more organized, with the DNA packaged within structures similar to eukaryotic nucleosomes.

Bacterial chromosomes tend to be tethered to the plasma membrane of the bacteria. In molecular biology application, this allows for its isolation from plasmid DNA by centrifugation of lysed bacteria and pelleting of the membranes (and the attached DNA). Prokaryotic chromosomes and plasmids are, like eukaryotic DNA, generally supercoiled. The DNA must first be released into its relaxed state for access for transcription, regulation, and replication.

Number of Chromosomes in Various Organisms

Eukaryotes

Human cells are diploid and have 22 different types of autosome, each present as two copies, and two sex chromosomes. This gives 46 chromosomes in total. Other organisms have more than two copies of their chromosomes, such as bread wheat, which is *hexaploid* and has six copies of seven different chromosomes – 42 chromosomes in total. Asexually reproducing species have one set of chromosomes, which are the same in all body cells. However, asexual species can be either haploid or diploid.

Sexually reproducing species have somatic cells (body cells), which are diploid [2n] having two sets of chromosomes, one from the mother and one from the father. Gametes, reproductive cells, are haploid [n]: They have one set of chromosomes. Gametes are produced by meiosis of a diploid germ line cell. During meiosis, the matching chromosomes of father and mother can exchange small parts of themselves (crossover), and thus create new chromosomes that are not inherited solely from either parent. When a male and a female gamete merge (fertilization), a new diploid organism is formed.

Some animal and plant species are polyploid [Xn]: They have more than two sets of homologous chromosomes. Plants important in agriculture such as tobacco or wheat are often polyploid, compared to their ancestral species. Wheat has a haploid number of seven chromosomes, still seen in some cultivars as well as the wild progenitors. The more-common pasta and bread wheats are polyploid, having 28 (tetraploid) and 42 (hexaploid) chromosomes, compared to the 14 (diploid) chromosomes in the wild wheat.

Prokaryotes

Prokaryote species generally have one copy of each major chromosome, but most cells can easily survive with multiple copies. For example, *Buchnera*, a symbiont of aphids has multiple copies of its chromosome, ranging from 10–400 copies per cell. However, in some large bacteria, such as *Epulopiscium fishelsoni* up to 100,000 copies of the chromosome can be present. Plasmids and plasmid-like small chromosomes are, as in eukaryotes, very variable in copy number. The number of plasmids in the cell is almost entirely determined by the rate of division of the plasmid – fast division causes high copy number, and vice versa.

Karyotype

In general, the karyotype is the characteristic chromosome complement of a eukaryote species. The preparation and study of karyotypes is part of cytogenetics. Although the replication and transcription of DNA is highly standardized in eukaryotes, *the same cannot be said for their karyotypes*, which are often highly variable. There may be variation between species in chromosome number and in detailed organization. In some cases, there is significant variation within species. Often there is:

1. variation between the two sexes
2. variation between the germ-line and soma (between gametes and the rest of the body)
3. variation between members of a population, due to balanced genetic polymorphism
4. geographical variation between races
5. mosaics or otherwise abnormal individuals.

Also, variation in karyotype may occur during development from the fertilised egg. The technique of determining the karyotype is usually called *karyotyping*. Cells can be locked part-way through division (in metaphase) in vitro (in a reaction vial) with colchicine. These cells are then stained, photographed, and arranged into a *karyogram*, with the set of chromosomes arranged, autosomes in order of length, and sex chromosomes (here X/Y) at the end:

Like many sexually reproducing species, humans have special gonosomes (sex chromosomes, in contrast to autosomes). These are XX in females and XY in males.

Historical Note

Investigation into the human karyotype took many years to settle the most basic question. How many chromosomes does a normal diploid human cell contain? In 1912, Hans von Winiwarter reported 47 chromosomes in spermatogonia and 48 in oogonia, concluding an

XX/XO sex determination mechanism. Painter in 1922 was not certain whether the diploid number of man is 46 or 48, at first favouring 46. He revised his opinion later from 46 to 48, and he correctly insisted on man's having an XX/XY system.

New techniques were needed to definitively solve the problem:

1. Using cells in culture
2. Pretreating cells in a hypotonic solution, which swells them and spreads the chromosomes
3. Arresting mitosis in metaphase by a solution of colchicine
4. Squashing the preparation on the slide forcing the chromosomes into a single plane
5. Cutting up a photomicrograph and arranging the result into an indisputable karyogram.

It took until the mid-1950s for it to become generally accepted that the human karyotype include only 46 chromosomes. Considering the techniques of Winiwarter and Painter, their results were quite remarkable. Chimpanzees (the closest living relatives to modern humans) have 48 chromosomes.

Chromosomal Aberrations

Chromosomal aberrations are disruptions in the normal chromosomal content of a cell, and are a major cause of genetic conditions in humans, such as Down syndrome. Some chromosome abnormalities do not cause disease in carriers, such as translocations, or chromosomal inversions, although they may lead to a higher chance of birthing a child with a chromosome disorder. Abnormal numbers of chromosomes or chromosome sets, aneuploidy, may be lethal or give rise to genetic disorders. Genetic counseling is offered for families that may carry a chromosome rearrangement.

The gain or loss of DNA from chromosomes can lead to a variety of genetic disorders. Human examples include:

- Cri du chat, which is caused by the deletion of part of the short arm of chromosome 5. "Cri du chat" means "cry of the cat" in French, and the condition was so-named because affected babies make high-pitched cries that sound like those of a cat. Affected individuals have wide-set eyes, a small head and jaw, moderate to severe mental health issues, and are very short.
- Down syndrome, usually is caused by an extra copy of chromosome 21 (trisomy 21). Characteristics include decreased

muscle tone, stockier build, asymmetrical skull, slanting eyes and mild to moderate developmental disability.

- Edwards syndrome, which is the second-most-common trisomy; Down syndrome is the most common. It is a trisomy of chromosome 18. Symptoms include motor retardation, developmental disability and numerous congenital anomalies causing serious health problems. Ninety percent die in infancy; however, those that live past their first birthday usually are quite healthy thereafter. They have a characteristic clenched hands and overlapping fingers.
- Idic15, abbreviation for Isodicentric 15 on chromosome 15; also called the following names due to various researches, but they all mean the same; IDIC(15), Inverted duplication 15, extra Marker, Inv dup 15, partial tetrasomy 15
- Jacobsen syndrome, also called the terminal 11q deletion disorder. This is a very rare disorder. Those affected have normal intelligence or mild developmental disability, with poor expressive language skills. Most have a bleeding disorder called Paris-Trousseau syndrome.
- Klinefelter's syndrome (XXY). Men with Klinefelter syndrome are usually sterile, and tend to have longer arms and legs and to be taller than their peers. Boys with the syndrome are often shy and quiet, and have a higher incidence of speech delay and dyslexia. During puberty, without testosterone treatment, some of them may develop gynecomastia.
- Patau Syndrome, also called D-Syndrome or trisomy-13. Symptoms are somewhat similar to those of trisomy-18, but they do not have the characteristic hand shape.
- Small supernumerary marker chromosome. This means there is an extra, abnormal chromosome. Features depend on the origin of the extra genetic material. Cat-eye syndrome and isodicentric chromosome 15 syndrome (or Idic15) are both caused by a supernumerary marker chromosome, as is Pallister-Killian syndrome.
- Triple-X syndrome (XXX). XXX girls tend to be tall and thin. They have a higher incidence of dyslexia.
- Turner syndrome (X instead of XX or XY). In Turner syndrome, female sexual characteristics are present but underdeveloped. People with Turner syndrome often have a short stature, low hairline, abnormal eye features and bone development and a "caved-in" appearance to the chest.

- XYY syndrome. XYY boys are usually taller than their siblings. Like XXY boys and XXX girls, they are somewhat more likely to have learning difficulties.
- Wolf-Hirschhorn syndrome, which is caused by partial deletion of the short arm of chromosome 4. It is characterized by severe growth retardation and severe to profound mental health issues.

Chromosomal mutations produce changes in whole chromosomes (more than one gene) or in the number of chromosomes present.

- Deletion – loss of part of a chromosome
- Duplication – extra copies of a part of a chromosome
- Inversion – reverse the direction of a part of a chromosome
- Translocation – part of a chromosome breaks off and attaches to another chromosome.

Most mutations are neutral – have little or no effect. Chromosomal aberrations are the changes in the structure of chromosomes. It has a great role in evolution. A detailed graphical display of all human chromosomes and the diseases annotated at the correct spot may be found at the Oak Ridge National Laboratory.

Chapter

9

Chemical Evolution of Cell

Chemical evolution describes chemical changes on the primitive Earth that gave rise to the first forms of life. The first living things on Earth were prokaryotes with a type of cell similar to present-day bacteria. Prokaryote fossils have been found in 3.4-million-year-old rock in the southern part of Africa, and in even older rocks in Australia, including some that appear to be photosynthetic. All forms of life are theorized to have evolved from the original prokaryotes, probably 3.5-4.0 billion years ago.

The chemical and physical conditions of the primitive Earth are invoked to explain the origin of life, which was preceded by chemical evolution of organic chemicals. Astronomers believe that 20-30 billion years ago, all matter was concentrated in a single mass, and that it blew apart with a "big bang." In time, a disk-shaped cloud of dust condensed and formed the Sun, and the peripheral matter formed its planets. Heat produced by compaction, radiation, and impacting meteorites melted Earth. Then, as the planet cooled, Earth's layers formed.

The first atmosphere was made up of hot hydrogen gas, too light to be held by Earth's gravity. Water vapour, carbon monoxide, carbon dioxide, nitrogen, and methane replaced the hydrogen atmosphere. As Earth cooled, water vapour condensed and torrential rains filled up its basins, thereby forming the seas. Also present were lightning, volcanic activity, and ultraviolet radiation. It was in this setting that life began.

According to one theory, chemical evolution occurred in four stages. In the first stage of chemical evolution, molecules in the primitive environment formed simple organic substances, such as amino acids. This concept was first proposed in 1936 in a book entitled, "The Origin of Life on Earth," written by the Russian scientist, Aleksandr Ivanovich Oparin.

He considered hydrogen, ammonia, water vapour, and methane to be components in the early atmosphere. Oxygen was lacking in this chemically-reducing environment. He stated that ultraviolet radiation from the Sun provided the energy for the transformation of these substances into organic molecules. Scientists today state that such spontaneous synthesis occurred only in the primitive environment. Abiogenesis became impossible when photosynthetic cells added oxygen to the atmosphere. The oxygen in the atmosphere gave rise to the ozone layer which then shielded Earth from ultraviolet radiation. Newer versions of this hypothesis contend that the primitive atmosphere also contained carbon monoxide, carbon dioxide, nitrogen, hydrogen sulfide, and hydrogen. Present-day volcanoes emit these substances.

In 1957, Stanley Miller and Harold Urey provided laboratory evidence that chemical evolution as described by Oparin could have occurred. Miller and Urey created an apparatus that simulated the primitive environment. They used a warmed flask of water for the ocean, and an atmosphere of water, hydrogen, ammonia and methane.

Sparks discharged into the artificial atmosphere represented lightning. A condenser cooled the atmosphere, causing rain that returned water and dissolved compounds back to the simulated sea. When Miller and Urey analysed the components of the solution after a week, they found various organic compounds had formed. These included some of the amino acids that compose the proteins of living things. Their results gave credence to the idea that simple substances in the warm primordial seas gave rise to the chemical building blocks of organisms.

In the second stage of chemical evolution, the simple organic molecules (such as amino acids) that formed and accumulated joined together into larger structures (such as proteins). The units linked to each other by the process of dehydration synthesis to form polymers. The problem is that the abiotic synthesis of polymers had to occur without the assistance of enzymes. In addition, these reactions give off water and would, therefore, not occur spontaneously in a watery environment. Sydney Fox of the University of Miami suggested that waves or rain in the primitive environment splashed organic monomers on fresh lava or hot rocks, which would have allowed polymers to form abiotically. When he tried to do this in his laboratory, Fox produced proteinoids—abiotically synthesized polypeptides.

The next step in chemical evolution suggests that polymers interacted with each other and organized into aggregates, known as protobionts. Protobionts are not capable of reproducing, but had other properties of living things. Scientists have successfully produced protobionts from organic molecules in the laboratory. In one study, proteinoids mixed with cool water assembled into droplets or microspheres that developed membranes on their surfaces. These are protobionts, with semipermeable and excitable membranes, similar to those found in cells.

In the final step of chemical evolution, protobionts developed the ability to reproduce and pass genetic information from one generation to the next. Some scientists theorize RNA to be the original hereditary molecule. Short polymers of RNA have been synthesized abiotically in the laboratory. In the 1980s, Thomas Cech and his associates at the University of Colorado at Boulder discovered that RNA molecules can function as enzymes in cells.

This implies that RNA molecules could have replicated in prebiotic cells without the use of protein enzymes. Variations of RNA molecules could have been produced by mutations and by errors during replication. Natural selection, operating on the different RNAs would have brought about subsequent evolutionary development. This would have fostered the survival of RNA sequences best suited to environmental parameters, such as temperature and salt concentration. As the protobionts grew and split, their RNA was passed on to offspring. In time, a diversity of prokaryote cells came into existence. Under the influence of natural selection, the prokaryotes could have given rise to the vast variety of life on Earth.

The evolutionary history of life on Earth traces the processes by which living and fossil organisms evolved. It stretches from the origin of life on Earth, thought to be over 3,500 million years ago, to the present day. The similarities between all present day organisms indicate the presence of a common ancestor from which all known species have diverged through the process of evolution.

Microbial mats of coexisting bacteria and archaea were the dominant form of life in the early Archean and many of the major steps in early evolution are thought to have taken place within them. The evolution of oxygenic photosynthesis, around 3,500 million years ago, eventually led to the oxygenation of the atmosphere, beginning around 2,400 million years ago.

The earliest evidence of eukaryotes (complex cells with organelles), dates from 1,850 million years ago, and while they may have been present earlier, their diversification accelerated when they started using oxygen in their metabolism. Later, around 1,700 million years ago, multicellular organisms began to appear, with differentiated cells performing specialised functions. The earliest land plants date back to around 450 million years ago, though evidence suggests that algal scum formed on the land as early as 1,200 million years ago. Land plants were so successful that they are thought to have contributed to the late Devonian extinction event. Invertebrate animals appear during the Vendian period, while vertebrates originated about 525 million years ago during the Cambrian explosion.

During the Permian period, synapsids, including the ancestors of mammals, dominated the land, but the Permian–Triassic extinction event 251 million years ago came close to wiping out all complex life. During the recovery from this catastrophe, archosaurs became the most abundant land vertebrates, displacing therapsids in the mid-Triassic. One archosaur group, the dinosaurs, dominated the Jurassic and Cretaceous periods, with the ancestors of mammals surviving only as small insectivores. After the Cretaceous–Tertiary extinction event 65 million years ago killed off the non-avian dinosaurs mammals increased rapidly in size and diversity. Such mass extinctions may have accelerated evolution by providing opportunities for new groups of organisms to diversify.

Fossil evidence indicates that flowering plants appeared and rapidly diversified in the Early Cretaceous, between 130 million years ago and 90 million years ago, probably helped by coevolution with pollinating insects. Flowering plants and marine phytoplankton are still the dominant producers of organic matter. Social insects appeared around the same time as flowering plants. Although they occupy only small parts of the insect "family tree", they now form over half the total mass of insects. Humans evolved from a lineage of upright-walking apes whose earliest fossils date from over 6 million years ago. Although early members of this lineage had chimpanzee-sized brains, there are signs of a steady increase in brain size after about 3 million years ago.

Earliest History of Earth

The oldest meteorite fragments found on Earth are about 4,540 million years old, this, coupled primarily with the dating of ancient lead deposits, has put the estimated age of Earth at around that time. About 40 million years later a planetoid struck the Earth, throwing into orbit the material that formed the Moon. Until recently the oldest rocks found on Earth were about 3,800 million years old, leading scientists to believe for decades that Earth's surface had been molten until then. Accordingly, they named this part of Earth's history the Hadean eon, whose name means "hellish". However analysis of zircons formed 4,400 to 4,000 million years ago indicates that Earth's crust solidified about 100 million years after the planet's formation and that the planet quickly acquired oceans and an atmosphere, which may have been capable of supporting life.

Evidence from the Moon indicates that from 4,000 to 3,800 million years ago it suffered a Late Heavy Bombardment by debris that was left over from the formation of the Solar system, and the Earth should have experienced an even heavier bombardment due to its stronger gravity. While there is no direct evidence of conditions on Earth 4,000 to 3,800 million years ago, there is no reason to think that the Earth was not also affected by this late heavy bombardment. This event may well have stripped away any previous atmosphere and oceans; in this case gases and water from comet impacts may have contributed to their replacement, although volcanic outgassing on Earth would have contributed at least half.

Earliest Evidence for Life on Earth

The earliest identified organisms were minute and relatively featureless, their fossils look like small rods, which are very difficult to tell apart from structures that arise through abiotic physical processes. The oldest undisputed evidence of life on Earth, interpreted as fossilized bacteria, dates to 3,000 million years ago. Other finds in rocks dated to about 3,500 million years ago have been interpreted as bacteria, with geochemical evidence also seeming to show the presence of life 3,800 million years ago.

However these analyses were closely scrutinized, and non-biological processes were found which could produce all of the "signatures of life" that had been reported. While this does not prove that the structures found had a non-biological origin, they cannot be taken as clear evidence for the presence of life. Currently, the oldest unchallenged evidence for life is geochemical signatures from rocks deposited 3,400 million years ago, although these statements have not been thoroughly examined by critics.

Origins of Life on Earth

Evolutionary tree showing the divergence of modern species from their common ancestor in the centre. The three domains are coloured, with bacteria blue, archaea green, and eukaryotes red. Biologists reason that all living organisms on Earth must share a single last universal ancestor, because it would be virtually impossible that two or more separate lineages could have independently developed the many complex biochemical mechanisms common to all living organisms.

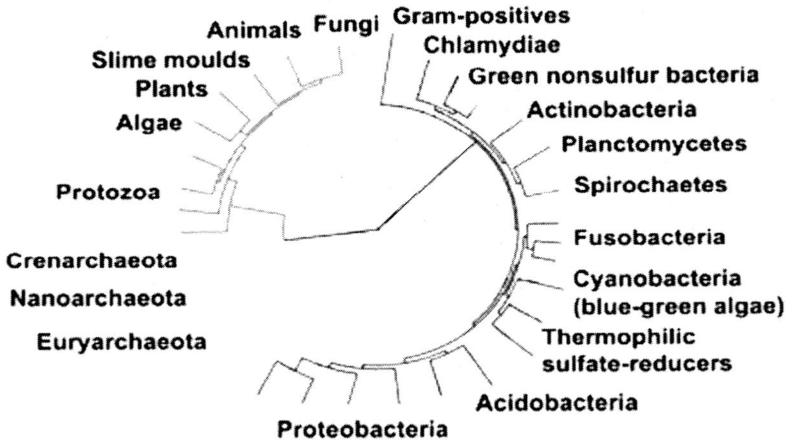

As previously mentioned the earliest organisms for which fossil evidence is available are bacteria, cells far too complex to have arisen directly from non-living materials. The lack of fossil or geochemical evidence for earlier organisms has left plenty of scope for hypotheses, which fall into two main groups: 1) that life arose spontaneously on Earth or 2) that it was "seeded" from elsewhere in the universe.

Life "Seeded" from Elsewhere

The idea that life on Earth was "seeded" from elsewhere in the universe dates back at least to the fifth century BCE. In the twentieth century it was proposed by the physical chemist Svante Arrhenius, by the astronomers Fred Hoyle and Chandra Wickramasinghe, and by molecular biologist Francis Crick and chemist Leslie Orgel. There are three main versions of the "seeded from elsewhere" hypothesis: from elsewhere in our Solar system via fragments knocked into space by a large meteor impact, in which case the only credible source is Mars; by alien visitors, possibly as a result of accidental contamination by micro-organisms that they brought with them; and from outside the Solar system but by natural means.

Experiments suggest that some micro-organisms can survive the shock of being catapulted into space and some can survive exposure to radiation for several days, but there is no proof that they can survive in space for much longer periods. Scientists are divided over the likelihood of life arising independently on Mars, or on other planets in our galaxy.

Independent Emergence on Earth

Life on Earth is based on carbon and water. Carbon provides stable frameworks for complex chemicals and can be easily extracted from the environment, especially from carbon dioxide. The only other element with similar chemical properties, silicon, forms much less stable structures and, because most of its compounds are solids, would be more difficult for organisms to extract.

Water is an excellent solvent and has two other useful properties: the fact that ice floats enables aquatic organisms to survive beneath it in winter; and its molecules have electrically negative and positive ends, which enables it to form a wider range of compounds than other solvents can. Other good solvents, such as ammonia, are liquid only at such low temperatures that chemical reactions may be too slow to sustain life, and lack water's other advantages. Organisms based on alternative biochemistry may however be possible on other planets.

Research on how life might have emerged unaided from non-living chemicals focuses on three possible starting points: self-replication, an organism's ability to produce offspring that are very similar to itself; metabolism, its ability to feed and repair itself; and external cell membranes, which allow food to enter and waste products to leave, but exclude unwanted substances. Research on abiogenesis still has a long way to go, since theoretical and empirical approaches are only beginning to make contact with each other.

Replication First: RNA World

The replicator in virtually all known life is deoxyribonucleic acid. DNA's structure and replication systems are far more complex than those of the original replicator. Even the simplest members of the three modern domains of life use DNA to record their "recipes" and a complex array of RNA and protein molecules to "read" these instructions and use them for growth, maintenance and self-replication. This system is far too complex to have emerged directly from non-living materials. The discovery that some RNA molecules can catalyse both their own replication and the construction of proteins led to the hypothesis of earlier life-forms based entirely on RNA.

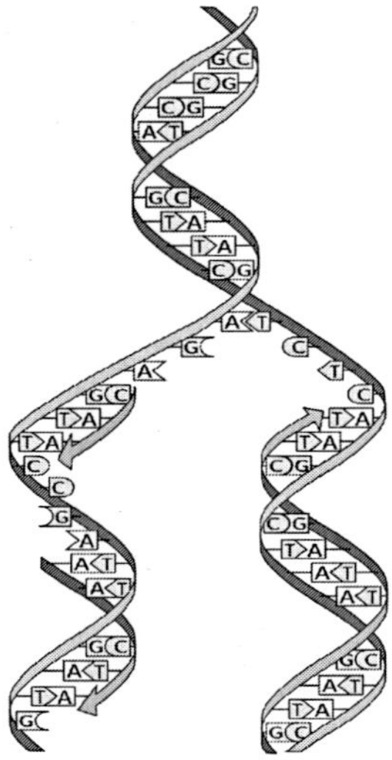

Figure: *The replicator in virtually all known life is deoxyribonucleic acid. DNA's structure and replication systems are far more complex than those of the original replicator.*

These ribozymes could have formed an RNA world in which there were individuals but no species, as mutations and horizontal gene transfers would have meant that the offspring in each generation were quite likely to have different genomes from those that their parents started with. RNA would later have been replaced by DNA, which is more stable and therefore can build longer genomes, expanding the range of capabilities a single organism can have. Ribozymes remain as the main components of ribosomes, modern cells' "protein factories". Although short self-replicating RNA molecules have been artificially produced in laboratories, doubts have been raised about where natural non-biological synthesis of RNA is possible. The earliest "ribozymes" may have been formed of simpler nucleic acids such as PNA, TNA or GNA, which would have been replaced later by RNA.

In 2003 it was proposed that porous metal sulfide precipitates would assist RNA synthesis at about 100 °C (212 °F) and ocean-bottom pressures near hydrothermal vents. In this hypothesis lipid membranes would be the last major cell components to appear and until then the proto-cells would be confined to the pores.

Metabolism First: Iron-sulfur World

A series of experiments starting in 1997 showed that early stages in the formation of proteins from inorganic materials including carbon monoxide and hydrogen sulfide could be achieved by using iron sulfide and nickel sulfide as catalysts. Most of the steps required temperatures of about 100 °C (212 °F) and moderate pressures, although one stage required 250 °C (482 °F) and a pressure equivalent to that found under 7 kilometres (4.3 mi) of rock. Hence it was suggested that self-sustaining synthesis of proteins could have occurred near hydrothermal vents.

Membranes First: Lipid World

It has been suggested that double-walled "bubbles" of lipids like those that form the external membranes of cells may have been an essential first step. Experiments that simulated the conditions of the early Earth have reported the formation of lipids, and these can spontaneously form liposomes, double-walled "bubbles", and then reproduce themselves. Although they are not intrinsically information-carriers as nucleic acids are, they would be subject to natural selection for longevity and reproduction. Nucleic acids such as RNA might then have formed more easily within the liposomes than they would have outside.

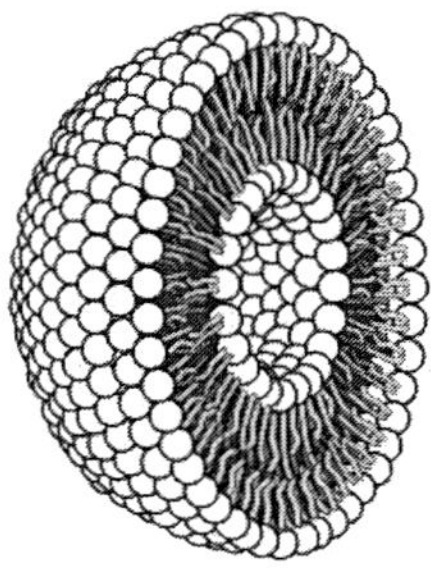

Figure: *Cross-section through a liposome.*

The Clay Theory

RNA is complex and there are doubts about whether it can be produced non-biologically in the wild. Some clays, notably montmorillonite, have properties that make them plausible accelerators for the emergence of an RNA world: they grow by self-replication of their crystalline pattern; they are subject to an analog of natural selection, as the clay "species" that grows fastest in a particular environment rapidly becomes dominant; and they can catalyse the formation of RNA molecules. Although this idea has not become the scientific consensus, it still has active supporters. Research in 2003 reported that montmorillonite could also accelerate the conversion of fatty acids into "bubbles", and that the "bubbles" could encapsulate RNA attached to the clay. These "bubbles" can then grow by absorbing additional lipids and then divide. The formation of the earliest cells may have been aided by similar processes. A similar hypothesis presents self-replicating iron-rich clays as the progenitors of nucleotides, lipids and amino acids.

RNA World Hypothesis

The RNA world hypothesis proposes that a world filled with life based on ribonucleic acid (RNA) predates the current world of life based on deoxyribonucleic acid (DNA) and protein. RNA, which can both store information like DNA and act as an enzyme like proteins, may have supported cellular or pre-cellular life. Some hypotheses as to the origin of life present RNA-based catalysis and information storage as the first step in the evolution of cellular life.

The RNA world is proposed to have evolved into the DNA and protein world of today. DNA, through its greater chemical stability, took over the role of data storage while protein, which is more flexible in catalysis through the great variety of amino acids, became the specialized catalytic molecules. The RNA world hypothesis suggests that RNA in modern cells, in particular rRNA (RNA in the ribosome which catalyzes protein production), is an evolutionary remnant of the RNA world.

History

The phrase "RNA World" was first used by Nobel laureate Walter Gilbert in 1986, in a commentary on recent observations of the catalytic properties of various forms of RNA. However, the idea of independent RNA life is older and can be found in Carl Woese's 1968 book *The Genetic Code*. In 1963, the molecular biologist Alexander Rich, of the Massachusetts Institute of Technology, had posited much the same idea in an article he contributed to a volume issued in honor of Nobel-laureate physiologist Albert Szent-Györgyi.

Properties of RNA

The properties of RNA make the idea of the RNA world hypothesis conceptually possible, although its plausibility as an explanation for the origin of life is debated. RNA is known to form efficient catalysts and its similarity to DNA makes its ability to store information clear. slightly different version of the hypothesis is that a different type of nucleic acid, termed *pre-RNA*, was the first one to emerge as a self-reproducing molecule, to be replaced by RNA only later.

This claim has been made due to (to date) failed attempts to efficiently synthesize activated RNA nucleotides which have the capability to undergo self-polymerization. Suggestions for 'simpler' nucleic acids include Peptide nucleic acid (PNA), Threose nucleic acid (TNA) or Glycerol nucleic acid (GNA). Despite their structural simplicity and possessing properties as good or greater than RNA, the chemically plausible generation of such monomers or oligomers under potentially prebiotic conditions has yet to be demonstrated.

RNA as an Enzyme

RNA enzymes, or ribozymes, are still found in today's DNA-based life and could be examples of living fossils. Ribozymes play vital roles, such as those in the ribosome, which is vital for protein synthesis. Many other ribozyme functions exist, for example: the Hammerhead ribozyme performs self-cleavage and an RNA polymerase ribozyme can autocatalyse its own synthesis. Among the enzymatic properties important for the beginning of life are:

- The ability to self-duplicate, or duplicate other RNA molecules. Relatively short RNA molecules that can duplicate others have been artificially produced in the lab. The shortest was 165-base long, though it has been estimated that only part of the bases were crucial for this function. One version, 189-base long, had fidelity of 98.9%, which would mean it would make an exact copy of an RNA molecule as long as itself in one of every eight copies. This 189 base pair ribozyme could polymerize a template of at most 14 nucleotides in length, which is too short for replication, but a promising lead for further investigation. The longest primer extension by a ribozyme polymerase was 20 bases.
- The ability to catalyse simple chemical reactions which would enhance the creation of molecules which are building blocks of RNA molecules—i.e., a strand of RNA which would make creating more strands of RNA easier. Relatively short RNA molecules with such abilities have been artificially formed in the lab.
- The ability to catalyse the formation of peptide bonds, in order to produce short peptides, or—eventually—full proteins. This

is done in modern cells by ribosomes, a complex of two large RNA molecules known as rRNA and many proteins. The two rRNA molecules are thought to be responsible for its enzymatic activity. A much shorter RNA molecule has been formed in lab with the ability to form peptide bonds, and it has been suggested that rRNA has evolved from a similar molecule. It has also been suggested that amino acids may have initially been complexed with RNA molecules as cofactors enhancing or diversifying their enzymatic capabilities, before evolving to the more complex peptides. mRNA may have evolved from such RNA molecules, and tRNA from RNA molecules which had catalysed amino acid transfer to them.

RNA in Information Storage

RNA is a very similar molecule to DNA, and only has two chemical differences. The overall structure of RNA and DNA are immensely similar—one strand of DNA and one of RNA can bind to form a double helical structure. This makes the storage of information in RNA possible in a very similar way to the storage of information in DNA.

Comparison of DNA and RNA Structure

The major difference between RNA and DNA is the presence of a hydroxyl group at the 2'-position of the ribose sugar in RNA. This group makes the molecule less stable—in flexible regions of an RNA molecule (i.e., where not constrained in a double helix), it can chemically attack the adjacent phosphodiester bond to cleave the phosphodiester backbone. The hydroxyl group also forces the ribose into the C3'-endo sugar conformation unlike the C2'-endo conformation of the deoxyribose sugar in DNA. This forces a RNA double helix into a slightly different conformation than B-DNA, resembling A-DNA more closely. RNA also uses a different set of bases than DNA—adenine, guanine, cytosine and uracil, instead of adenine, guanine, cytosine and thymine. Chemically, uracil is similar to thymine, differing only by a methyl group, and its production requires less energy. In terms of base pairing this has no effect, adenine will readily bind uracil or thymine. Uracil is, however, one product of damage to cytosine making RNA particularly susceptible to mutations which replace a GC base pair with a GU (wobble) or AU base pair.

Limitations of Information Storage in RNA

Storing large amounts of information in RNA is not easy. The chemical properties of RNA make large RNA molecules inherently fragile, and they can easily be broken down into their constituent nucleotides through hydrolysis. The aromatic bases also absorb strongly in the ultraviolet region, and would have been susceptible to damage and breakdown by background radiation. These limitations do not make use of RNA as an information storage system impossible, simply energy intensive (to repair or replace damaged RNA molecules) and mutation prone. While this makes it unsuitable for current 'DNA optimised' life, it may have been suitable for primitive life.

RNA as a Regulator

Riboswitches have been found to act as regulators of gene expression, particularly in bacteria, but also in plants and archaea. It has been suggested that these originated in an RNA based world.

Support and Difficulties

The RNA World hypothesis is supported by RNA's ability to store, transmit, and duplicate genetic information, as DNA does. RNA can also act as a ribozyme, a special type of enzyme. Because it can reproduce on its own, performing the tasks of both DNA and proteins (enzymes), RNA is believed to have once been capable of independent life. Further, while nucleotides were not found in Miller-Urey's origins of life experiments, their formation in prebiotically plausible conditions was reported recently; the purine base known as adenine is merely a pentamer of hydrogen cyanide. Experiments with basic ribozymes, like the viral RNA Qβ, have shown that simple self-replicating RNA structures can withstand even strong selective pressures (e.g., opposite-chirality chain terminators).

Additionally, in the past a given RNA molecule might have survived longer than it can today. Ultraviolet light can cause RNA to polymerize while at the same time breaking down other types of organic molecules that could have the potential of causing the breakdown of RNA, suggesting that RNA may have been a relatively common substance on early Earth. This aspect of the theory is still untested and is based on a constant concentration of sugar-phosphate molecules.

Since there were no known chemical pathways for the abiogenic synthesis of nucleotides from pyrimidine nucleobases cytosine and uracil under prebiotic conditions, it is thought by some that nucleic acids did not contain these nucleobases seen in life's nucleic acids. The nucleoside cytosine has a half-life in isolation of 19 days at 100 °C (212 °F) and 17,000 years in freezing water, which has been argued to be too short on the geologic time scale for accumulation. Others have questioned whether ribose and other backbone sugars could be stable enough to be found in the original genetic material, and have raised the issue that ribose must all be the same enantiomer as any nucleotide of the wrong chirality acts as a chain terminator.

Pyrimidine ribonucleosides and their respective nucleotides have been prebiotically synthesised by a sequence of reactions which by-pass the free sugars, and is assembled in a stepwise fashion by going against the dogma that nitrogenous and oxygenous chemistries should be avoided. In a series of publications, The *Sutherland Group* at the School of Chemistry, University of Manchester have demonstrated high yielding routes to cytidine and uridine ribonucleotides built from small 2 and 3 carbon fragments such as glycolaldehyde, glyceraldehyde or glyceraldehyde-3-phosphate, cyanamide and cyanoacetylene. One of the steps in this sequence allows the isolation of enantiopure ribose aminooxazoline if the enantiomeric excess of glyceraldehyde is 60 % or greater, of possible interest towards biological homochirality.

This can be viewed as a prebiotic purification step, where the said compound spontaneously crystallised out from a mixture of the other pentose aminooxazolines. Aminooxazolines can react with cyanoacetylene in a mild and highly efficient manner, controlled by inorganic phosphate, to give the cytidine ribonucleotides. Photoanomerization with UV light allows for inversion about the 1' anomeric centre to give the correct beta stereochemistry, one problem with this chemistry is the selective phosphorylation of alpha-cytidine at the 2' position. However, in 2009 they showed that the same simple building blocks allow access, via phosphate controlled nucleobase elaboration, to 2',3'-cyclic pyrimidine nucleotides directly, which are known to be able to polymerise into RNA.

This was hailed as strong evidence for the RNA world. The paper also highlighted the possibility for the photo-sanitization of the pyrimidine-2',3'-cyclic phosphates. A potential weakness of these routes is the generation of enantioenriched glyceraldehyde, or its 3-phosphate derivative (glyceraldehyde prefers to exist as its keto tautomer dihydroxyacetone).

Details of the RNA World

Mechanism for Prebiotic RNA Synthesis

Nucleotides are the fundamental molecules that combine in series to form RNA. They consist of a nitrogenous base attached to a sugar-phosphate backbone. RNA is made of long stretches of specific nucleotides arranged so that their sequence of bases carries information. The RNA world hypothesis holds that in the primordial soup/primordial sandwich, there existed free-floating nucleotides. These nucleotides regularly formed bonds with one another, which often broke because the change in energy was so low. However, certain sequences of base pairs have catalytic properties that lower the energy of their chain being created, causing them to stay together for longer periods of time. As each chain grew longer, it attracted more matching nucleotides faster, causing chains to now form faster than they were breaking down.

These chains are proposed as the first, primitive forms of life. In an RNA world, different forms of RNA compete with each other for free nucleotides and are subject to natural selection. The most efficient molecules of RNA, the ones able to efficiently catalyse their own reproduction, survived and evolved, forming modern RNA. Such an RNA enzyme, capable of self replication in about an hour has been identified. It was produced by molecular competition (*in vitro* evolution) of candidate enzyme mixtures.

Competition between RNA may have favoured the emergence of cooperation between different RNA chains, opening the way for the formation of the first proto-cell. Eventually, RNA chains randomly developed with catalytic properties that help amino acids bind together (a process called peptide-bonding). These amino acids could then assist with RNA synthesis, giving those RNA chains that could serve as ribozymes the selective advantage. The ability to catalyse one step in protein synthesis, aminoacylation of RNA, has been demonstrated in a short (five-nucleotide) segment of RNA.

Further Developments

Patrick Forterre has been working on a novel hypothesis, that viruses were instrumental in the transition from RNA to DNA and the evolution of Bacteria, Archaea, and Eukaryota. He believes the last common ancestor was RNA-based and evolved RNA viruses. Some of the viruses evolved into DNA viruses to protect their genes from attack. Through the process of viral infection into hosts the three domains of life evolved. Another interesting proposal is the idea that RNA synthesis might have been driven by temperature gradients, in the process of thermosynthesis. Recently Dr.Atul Kumar has been the first to demonstrate that single nucleotides have the ability to catalyse organic reactions. This single-nucleotide catalysis has immense impact on many fields of science such as chemistry, biochemistry, and prebiotic studies, especially the RNA world and DNA world hypothesis for understanding the origin of life on Earth..

Alternative Hypotheses

A different version of the same hypothesis is "Pre-RNA world", where a different nucleic acid is proposed to pre-date RNA. One of the alternate nucleic acid hypotheses is the PNA world hypothesis involving peptide nucleic acid, PNA. PNA is more stable than RNA, but its ability to be generated under prebiological conditions has yet to be demonstrated experimentally.

Threose nucleic acid (TNA) has also been proposed as a starting point, as has glycol nucleic acid (GNA), and like PNA, also lack experimental evidence for their respective abiogenesis. A different—or complementary—alternative to the assembly of RNA is proposed in the PAH world hypothesis. The iron-sulfur world theory proposes that simple metabolic processes developed before genetic materials did, and these energy-producing cycles catalysed the production of genes. Yet another alternative theory to the RNA world hypothesis is the panspermia hypothesis. It discusses the possibility that the earliest life on this planet was carried here from somewhere else in the galaxy, possibly on meteorites similar to the Murchison meteorite. This does not invalidate the concept of an RNA world, but posits that this world was not Earth but rather another, probably older planet.

Implications of the RNA World

The RNA world hypothesis, if true, has important implications for the very definition of life. For the majority of the time following the elucidation of the structure of DNA by Watson and Crick, life was considered as being largely defined in terms of DNA and proteins: DNA and proteins seemed to be the dominant macromolecules in the living cell, with RNA serving only to aid in creating proteins from the DNA blueprint.

The RNA world hypothesis places RNA at centre-stage when life originated. This has been accompanied by many studies in the last ten years demonstrating important aspects of RNA function that were not previously known, and support the idea of a critical role for RNA in the functionality of life. In 2001, the RNA world hypothesis was given a major boost with the deciphering of the 3-dimensional structure of the ribosome, which revealed the key catalytic sites of ribosomes to be composed of RNA and for the proteins to hold no major structural role, and be of peripheral functional importance. Specifically, the formation of the peptide bond, the reaction that binds amino acids together into proteins, is now known to be catalysed by an adenine residue in the rRNA: the ribosome is a ribozyme. This finding suggests that RNA molecules were most likely capable of generating the first proteins. Other interesting discoveries demonstrating a role for RNA beyond a simple message or transfer molecule include the importance of small nuclear ribonucleoproteins (SnRNPs) in the processing of pre-mRNA and RNA editing and reverse transcription from RNA in Eukaryotes in the maintenance of telomeres in the telomerase reaction.

Environmental and Evolutionary Impact of Microbial Mats

Microbial mats are multi-layered, multi-species colonies of bacteria and other organisms that are generally only a few millimeters thick, but still contain a wide range of chemical environments, each of which favors a different set of micro-organisms. To some extent each mat forms its own food chain, as the by-products of each group of micro-organisms generally serve as "food" for adjacent groups. Stromatolites are stubby pillars built as microbes in mats slowly migrate upwards to avoid being smothered by sediment deposited on them by water. There has been vigorous debate about the validity of alleged fossils from before 3,000 million years ago, with critics arguing that so-called stromatolites could have been formed by non-biological processes. In 2006 another find of stromatolites was reported from the same part of Australia as previous ones, in rocks dated to 3,500 million years ago.

In modern underwater mats the top layer often consists of photosynthesizing cyanobacteria which create an oxygen-rich environment, while the bottom layer is oxygen-free and often dominated by hydrogen sulfide emitted by the organisms living there. It is estimated that the appearance of oxygenic photosynthesis by bacteria in mats increased biological productivity by a factor of between 100 and 1,000.

The reducing agent used by oxygenic photosynthesis is water, which is much more plentiful than the geologically-produced reducing agents required by the earlier non-oxygenic photosynthesis. From this point onwards life itself produced significantly more of the resources it needed than did geochemical processes. Oxygen is toxic to organisms that are not adapted to it, but greatly increases the metabolic efficiency of oxygen-adapted organisms. Oxygen became a significant component of Earth's atmosphere about 2,400 million years ago. Although eukaryotes may have been present much earlier, the oxygenation of the atmosphere was a prerequisite for the evolution of the most complex eukaryotic cells, from which all multicellular organisms are built. The boundary between oxygen-rich and oxygen-free layers in microbial mats would have moved upwards when photosynthesis shut down overnight, and then downwards as it resumed on the next day. This would have created selection pressure for organisms in this intermediate zone to acquire the ability to tolerate and then to use oxygen, possibly via endosymbiosis, where one organism lives inside another and both of them benefit from their association. Cyanobacteria have the most complete biochemical "toolkits" of all the mat-forming organisms. Hence they are the most self-sufficient of the mat organisms and were well-adapted to strike out on their own both as floating mats and as the first of the phytoplankton, providing the basis of most marine food chains.

Diversification of Eukaryotes

Eukaryotes may have been present long before the oxygenation of the atmosphere, but most modern eukaryotes require oxygen, which their mitochondria use to fuel the production of ATP, the internal energy supply of all known cells. In the 1970s it was proposed and, after much debate, widely accepted that eukaryotes emerged as a result of a sequence of endosymbioses between "procaryotes". For example: a predatory micro-organism invaded a large procaryote, probably an archaean, but the attack was neutralized, and the attacker took up residence and evolved into the first of the mitochondria; one of these chimeras later tried to swallow a photosynthesizing cyanobacterium, but the victim survived inside the attacker and the new combination became the ancestor of plants; and so on. After each endosymbiosis began, the partners would have eliminated unproductive duplication of genetic functions by re-arranging their genomes, a process which sometimes involved transfer of genes between them. Another hypothesis proposes that mitochondria were originally sulfur-or hydrogen-metabolising endosymbionts, and became oxygen-consumers later. On the other hand mitochondria might have been part of eukaryotes' original equipment.

There is a debate about when eukaryotes first appeared: the presence of steranes in Australian shales may indicate that eukaryotes were present 2,700 million years ago; however an analysis in 2008 concluded that these chemicals infiltrated the rocks less than 2,200 million years ago and prove nothing about the origins of eukaryotes. Fossils of the alga *Grypania* have been reported in 1,850 million-year-old rocks (originally dated to 2,100 million years ago but later revised), and indicates that eukaryotes with organelles had already evolved. A diverse collection of fossil algae were found in rocks dated between 1,500 million years ago and 1,400 million years ago. The earliest known fossils of fungi date from 1,430 million years ago.

Multicellular Organisms and Sexual Reproduction

Multicellularity

The simplest definitions of "multicellular", for example "having multiple cells", could include colonial cyanobacteria like Nostoc. Even a professional biologist's definition such as "having the same genome but different types of cell" would still include some genera of the green alga Volvox, which have cells that specialize in reproduction. Multicellularity evolved independently in organisms as diverse as sponges and other animals, fungi, plants, brown algae, cyanobacteria, slime moulds and myxobacteria. For the sake of brevity this article focuses on the organisms that show the greatest specialization of cells and variety of cell types, although this approach to the evolution of complexity could be regarded as "rather anthropocentric".

The initial advantages of multicellularity may have included: increased resistance to predators, many of which attacked by engulfing; the ability to resist currents by attaching to a firm surface; the ability to reach upwards to filter-feed or to obtain sunlight for photosynthesis; the ability to create an internal environment that gives protection against the external one; and even the opportunity for a group of cells to behave "intelligently" by sharing information.

These features would also have provided opportunities for other organisms to diversify, by creating more varied environments than flat microbial mats could.

Multicellularity with differentiated cells is beneficial to the organism as a whole but disadvantageous from the point of view of individual cells, most of which lose the opportunity to reproduce themselves.

In an asexual multicellular organism, rogue cells which retain the ability to reproduce may take over and reduce the organism to a mass of undifferentiated cells. Sexual reproduction eliminates such rogue cells from the next generation and therefore appears to be a prerequisite for complex multicellularity.

The available evidence indicates that eukaryotes evolved much earlier but remained inconspicuous until a rapid diversification around

1,000 million years ago. The only respect in which eukaryotes clearly surpass bacteria and archaea is their capacity for variety of forms, and sexual reproduction enabled eukaryotes to exploit that advantage by producing organisms with multiple cells that differed in form and function.

Evolution of Sexual Reproduction

The defining characteristic of sexual reproduction is recombination, in which each of the offspring receives 50% of its genetic inheritance from each of the parents. Bacteria also exchange DNA by bacterial conjugation, the benefits of which include resistance to antibiotics and other toxins, and the ability to utilize new metabolites. However conjugation is not a means of reproduction, and is not limited to members of the same species – there are cases where bacteria transfer DNA to plants and animals.

The disadvantages of sexual reproduction are well-known: the genetic reshuffle of recombination may break up favourable combinations of genes; and since males do not directly increase the number of offspring in the next generation, an asexual population can out-breed and displace in as little as 50 generations a sexual population that is equal in every other respect. Nevertheless the great majority of animals, plants, fungi and protists reproduce sexually. There is strong evidence that sexual reproduction arose early in the history of eukaryotes and that the genes controlling it have changed very little since then. How sexual reproduction evolved and survived is an unsolved puzzle.

The Red Queen Hypothesis suggests that sexual reproduction provides protection against parasites, because it is easier for parasites to evolve means of overcoming the defenses of genetically identical clones than those of sexual species that present moving targets, and there is some experimental evidence for this. However there is still doubt about whether it would explain the survival of sexual species if multiple similar clone species were present, as one of the clones may survive the attacks of parasites for long enough to out-breed the sexual species.

The Mutation Deterministic Hypothesis assumes that each organism has more than one harmful mutation and the combined effects of these mutations are more harmful than the sum of the harm done by each individual mutation. If so, sexual recombination of genes will reduce the harm that bad mutations do to offspring and at the same time eliminate some bad mutations from the gene pool by isolating them in individuals that perish quickly because they have an above-average number of bad mutations. However the evidence suggests that the MDH's assumptions are shaky, because many species have on average less than one harmful mutation per individual and no species that has been investigated shows evidence of synergy between harmful mutations.

The random nature of recombination causes the relative abundance of alternative traits to vary from one generation to another. This genetic drift is insufficient on its own to make sexual reproduction advantageous, but a combination of genetic drift and natural selection may be sufficient. When chance produces combinations of good traits, natural selection gives a large advantage to lineages in which these traits become genetically linked. On the other hand the benefits of good traits are neutralized if they appear along with bad traits. Sexual recombination gives good traits the opportunities to become linked with other good traits, and mathematical models suggest this may be more than enough to offset the disadvantages of sexual reproduction. Other combinations of hypotheses that are inadequate on their own are also being examined.

The following hypotheses attempt to explain how and why sex evolved:

- It may have enabled organisms to repair genetic damage. The most primitive form of sex may have been one organism repairing damaged DNA by replicating an undamaged strand from a similar organism.
- Sexual reproduction may have originated from selfish parasitic genetic elements propagating themselves by transfer to new hosts.
- It may have evolved from cannibalism, where some of the victim's DNA was incorporated into the cannibal organism.
- Sexual reproduction may have evolved from ancient haloarchaea through a combination of jumping genes, and swapping plasmids.
- Or it may have evolved as a form of vaccination in which infected hosts exchanged weakened symbiotic copies of parasitic DNA as protection against more virulent versions. The meiosis stage of sexual reproduction may then have evolved as a way of removing the symbiotes.

Bacteria also exchange DNA by bacterial conjugation, the benefits of which include resistance to antibiotics and other toxins, and the ability to utilize new metabolites. However conjugation is not a means of reproduction and is not limited to members of the same species, and there are cases where bacteria transfer DNA to plants and animals. Nevertheless it may be an example of the "selfish genetic element" hypothesis, as it transfers DNA by means of such a "selfish gene", the F-plasmid.

Fossil Evidence for Multicellularity and Sexual Reproduction

The Francevillian Group Fossil, dated to 2,100 million years ago, is the earliest known fossil organism that is clearly multicellular. They may have had differentiated cells. Another early multicellular fossil, *Qingshania*, dated to 1,700 million years ago, appears to consist of virtually identical cells. The red alga called *Bangiomorpha*, dated at 1,200 million years ago, is the earliest known organism which certainly has differentiated, specialized cells, and is also the oldest known sexually-reproducing organism. The 1,430 million-year-old fossils interpreted as fungi appear to have been multicellular with differentiated cells. The "string of beads" organism *Horodyskia*, found in rocks dated from 1,500 million years ago to 900 million years ago, may have been an early metazoan; however it has also been interpreted as a colonial foraminiferan.

Emergence of Animals

The Ediacara biota, which flourished for the last 40 million years before the start of the Cambrian, were the first animals more than a very few centimetres long. Many were flat and had a "quilted" appearance, and seemed so strange that there was a proposal to classify them as a separate kingdom, Vendozoa. Others, however, been interpreted as early molluscs (*Kimberella*), echinoderms (*Arkarua*), and arthropods (*Spriggina*, *Parvancorina*). There is still debate about the classification of these specimens, mainly because the diagnostic features which allow taxonomists to classify more recent organisms, such as similarities to living organisms, are generally absent in the Ediacarans. However there seems little doubt that *Kimberella* was at least a triploblastic bilaterian animal, in other words significantly more complex than cnidarians.

The small shelly fauna are a very mixed collection of fossils found between the Late Ediacaran and Mid Cambrian periods. The earliest, *Cloudina*, shows signs of successful defence against predation and may indicate the start of an evolutionary arms race. Some tiny Early Cambrian shells almost certainly belonged to molluscs, while the owners of some "armor plates", *Halkieria* and *Microdictyon*, were eventually identified when more complete specimens were found in

Cambrian lagerstätten that preserved soft-bodied animals. In the 1970s there was already a debate about whether the emergence of the modern phyla was "explosive" or gradual but hidden by the shortage of Pre-Cambrian animal fossils.

A re-analysis of fossils from the Burgess Shale lagerstätte increased interest in the issue when it revealed animals, such as *Opabinia*, which did not fit into any known phylum. At the time these were interpreted as evidence that the modern phyla had evolved very rapidly in the "Cambrian explosion" and that the Burgess Shale's "weird wonders" showed that the Early Cambrian was a uniquely experimental period of animal evolution.

Later discoveries of similar animals and the development of new theoretical approaches led to the conclusion that many of the "weird wonders" were evolutionary "aunts" or "cousins" of modern groups – for example that *Opabinia* was a member of the lobopods, a group which includes the ancestors of the arthropods, and that it may have been closely related to the modern tardigrades. Nevertheless there is still much debate about whether the Cambrian explosion was really explosive and, if so, how and why it happened and why it appears unique in the history of animals.

Most of the animals at the heart of the Cambrian explosion debate are protostomes, one of the two main groups of complex animals. One deuterostome group, the echinoderms, many of which have hard calcite "shells", are fairly common from the Early Cambrian small shelly fauna onwards. Other deuterostome groups are soft-bodied, and most of the significant Cambrian deuterostome fossils come from the Chengjiang fauna, a lagerstätte in China. The Chengjiang fossils *Haikouichthys* and *Myllokunmingia* appear to be true vertebrates, and *Haikouichthys* had distinct vertebrae, which may have been slightly mineralized. Vertebrates with jaws, such as the Acanthodians, first appeared in the Late Ordovician.

Colonization of Land

Adaptation to life on land is a major challenge: all land organisms need to avoid drying-out and all those above microscopic size have to resist gravity; respiration and gas exchange systems have to change; reproductive systems cannot depend on water to carry eggs and sperm towards each other. Although the earliest good evidence of land plants and animals dates back to the Ordovician period (488 to 444 million years ago), modern land ecosystems only appeared in the late Devonian, about 385 to 359 million years ago.

Evolution of Soil

Before the colonization of land, soil, a combination of mineral particles and decomposed organic matter, did not exist. Land surfaces would have been either bare rock or unstable sand produced by weathering. Water and any nutrients in it would have drained away very quickly.

Films of cyanobacteria, which are not plants but use the same photosynthesis mechanisms, have been found in modern deserts, and only in areas that are unsuitable for vascular plants. This suggests that microbial mats may have been the first organisms to colonize dry land, possibly in the Precambrian. Mat-forming cyanobacteria could have gradually evolved resistance to desiccation as they spread from the seas to tidal zones and then to land.

Lichens, which are symbiotic combinations of a fungus (almost always an ascomycete) and one or more photosynthesizers (green algae or cyanobacteria), are also important colonizers of lifeless environments, and their ability to break down rocks contributes to soil formation in situations where plants cannot survive. The earliest known ascomycete fossils date from 423 to 419 million years ago in the Silurian.

Soil formation would have been very slow until the appearance of burrowing animals, which mix the mineral and organic components of soil and whose feces are a major source of the organic components. Burrows have been found in Ordovician sediments, and are attributed to annelids ("worms") or arthropods.

Plants and the Late Devonian Wood Crisis

In aquatic algae, almost all cells are capable of photosynthesies and are nearly independent. Life on land required plants to become internally more complex and specialized: photosynthesis was most efficient at the top; roots were required in order to extract water from the ground; the parts in between became supports and transport systems for water and nutrients.

Spores of land plants, possibly rather like liverworts, have been found in Mid Ordovician rocks dated to about 476 million years ago. In Mid Silurian rocks 430 million years ago there are fossils of actual plants including clubmosses such as *Baragwanathia*; most were under 10 centimetres (3.9 in) high, and some appear closely related to vascular plants, the group that includes trees. By the Late Devonian 370 million years ago, trees such as *Archaeopteris* were so abundant that they changed river systems from mostly braided to mostly meandering, because their roots bound the soil firmly. In fact they caused a "Late Devonian wood crisis", because:

- They removed more carbon dioxide from the atmosphere, reducing the greenhouse effect and thus causing an ice age in the Carboniferous period. In later ecosystems the carbon dioxide "locked up" in wood is returned to the atmosphere by decomposition of dead wood. However the earliest fossil evidence of fungi that can decompose wood also comes from the Late Devonian.
- The increasing depth of plants' roots led to more washing of nutrients into rivers and seas by rain. This caused algal blooms whose high consumption of oxygen caused anoxic events in deeper waters, increasing the extinction rate among deep-water animals.

Land Invertebrates

Animals had to change their feeding and excretory systems, and most land animals developed internal fertilization of their eggs. The difference in refractive index between water and air required changes in their eyes. On the other hand in some ways movement and breathing became easier, and the better transmission of high-frequency sounds in air encouraged the development of hearing.

Some trace fossils from the Cambrian-Ordovician boundary about 490 million years ago are interpreted as the tracks of large amphibious arthropods on coastal sand dunes, and may have been made by euthycarcinoids, which are thought to be evolutionary "aunts" of myriapods. Other trace fossils from the Late Ordovician a little over 445 million years ago probably represent land invertebrates, and there is clear evidence of numerous arthropods on coasts and alluvial plains shortly before the Silurian-Devonian boundary, about 415 million years ago, including signs that some arthropods ate plants. Arthropods were well pre-adapted to colonise land, because their existing jointed exoskeletons provided protection against desiccation, support against gravity and a means of locomotion that was not dependent on water.

The fossil record of other major invertebrate groups on land is poor: none at all for non-parasitic flatworms, nematodes or nemerteans; some parasitic nematodes have been fossilized in amber; annelid worm fossils are known from the Carboniferous, but they may still have been aquatic animals; the earliest fossils of gastropods on land date from the Late Carboniferous, and this group may have had to wait until leaf litter became abundant enough to provide the moist conditions they need. The earliest confirmed fossils of flying insects date from the Late Carboniferous, but it is thought that insects developed the ability to fly in the Early Carboniferous or even Late Devonian. This gave them a wider range of ecological niches for feeding and breeding, and a means of escape from predators and from unfavourable changes in the environment. About 99% of modern insect species fly or are descendants of flying species.

Land Vertebrates

Tetrapods, vertebrates with four limbs, evolved from other rhipidistians over a relatively short timespan during the Late Devonian, between 370 million years ago and 360 million years ago. From the 1950s to the early 1980s it was thought that tetrapods evolved from fish that had already acquired the ability to crawl on land, possibly in order to go from a pool that was drying out to one that was deeper.

However in 1987 nearly-complete fossils of *Acanthostega* from about 363 million years ago showed that this Late Devonian transitional animal had legs and both lungs and gills, but could never have survived on land: its limbs and its wrist and ankle joints were too weak to bear its weight; its ribs were too short to prevent its lungs from being squeezed flat by its weight; its fish-like tail fin would have been damaged by dragging on the ground. The current hypothesis is that *Acanthostega*, which was about 1 metre (3.3 ft) long, was a wholly aquatic predator that hunted in shallow water. Its skeleton differed from that of most fish, in ways that enabled it to raise its head to breathe air while its body remained submerged, including: its jaws show modifications that would have enabled it to gulp air; the bones at the back of its skull are locked together, providing strong attachment points for muscles that raised its head; the head is not joined to the shoulder girdle and it has a distinct neck.

The Devonian proliferation of land plants may help to explain why air-breathing would have been an advantage: leaves falling into streams and rivers would have encouraged the growth of aquatic vegetation; this would have attracted grazing invertebrates and small fish that preyed on them; they would have been attractive prey but the environment was unsuitable for the big marine predatory fish; air-breathing would have been necessary because these waters would have been short of oxygen, since warm water holds less dissolved oxygen than cooler marine water and since the decomposition of vegetation would have used some of the oxygen.

Later discoveries revealed earlier transitional forms between *Acanthostega* and completely fish-like animals. Unfortunately there is then a gap of about 30 million years between the fossils of ancestral tetrapods and Mid Carboniferous fossils of vertebrates that look well-adapted for life on land. Some of these look like early relatives of modern amphibians, most of which need to keep their skins moist and to lay their eggs in water, while others are accepted as early relatives of the amniotes, whose water-proof skins and eggs enable them to live and breed far from water.

Dinosaurs, Birds and Mammals

Amniotes, whose eggs can survive in dry environments, probably evolved in the Late Carboniferous period, between 330 million years ago and 314 million years ago. The earliest fossils of the two surviving amniote groups, synapsids and sauropsids, date from around 313 million years ago. The synapsid pelycosaurs and their descendants the therapsids are the most common land vertebrates in the best-known Permian fossil beds, between 229 million years ago and 251 million years ago. However at the time these were all in temperate zones at middle latitudes, and there is evidence that hotter, drier environments nearer the Equator were dominated by sauropsids and amphibians.

The Permian-Triassic extinction wiped out almost all land vertebrates, as well as the great majority of other life. During the slow recovery from this catastrophe, estimated to be 30M years, a previously obscure sauropsid group became the most abundant and diverse terrestrial vertebrates: a few fossils of archosauriformes ("shaped like archosaurs") have been found in Late Permian rocks, but by the Mid Triassic archosaurs were the dominant land vertebrates. Dinosaurs distinguished themselves from other archosaurs in the Late Triassic, and became the dominant land vertebrates of the Jurassic and Cretaceous periods, between 199 million years ago and 65 million years ago.

During the Late Jurassic, birds evolved from small, predatory theropod dinosaurs. The first birds inherited teeth and long, bony tails from their dinosaur ancestors, but some developed horny, toothless beaks by the very Late Jurassic and short pygostyle tails by the Early Cretaceous. While the archosaurs and dinosaurs were becoming more dominant in the Triassic, the mammaliform successors of the therapsids could only survive as small, mainly nocturnal insectivores. This apparent set-back may actually have promoted the evolution of mammals, for example nocturnal life may have accelerated the development of endothermy ("warm-bloodedness") and hair or fur. By 195 million years ago in the Early Jurassic there were animals that were very nearly mammals. Unfortunately there is a gap in the fossil record throughout the Mid Jurassic.

However fossil teeth discovered in Madagascar indicate that true mammals existed at least 167 million years ago. After dominating land vertebrate niches for about 150 million years, the dinosaurs perished 65 million years ago in the Cretaceous–Tertiary extinction along with many other groups of organisms. Mammals throughout the time of the dinosaurs had been restricted to a narrow range of taxa, sizes and shapes, but increased rapidly in size and diversity after the extinction, with bats taking to the air within 13 million years, and cetaceans to the sea within 15 million years.

Flowering Plants

The 250,000 to 400,000 species of flowering plants outnumber all other ground plants combined, and are the dominant vegetation in most terrestrial ecosystems. There is fossil evidence that flowering plants diversified rapidly in the Early Cretaceous, between 130 million years ago and 90 million years ago, and that their rise was associated with that of pollinating insects. Among modern flowering plants Magnolias are thought to be close to the common ancestor of the group. However paleontologists have not succeeded in identifying the earliest stages in the evolution of flowering plants.

Social Insects

The social insects are remarkable because the great majority of individuals in each colony are sterile. This appears contrary to basic concepts of evolution such as natural selection and the selfish gene. In fact there are very few eusocial insect species: only 15 out of approximately 2,600 living families of insects contain eusocial species, and it seems that eusociality has evolved independently only 12 times among arthropods, although some eusocial lineages have diversified into several families. Nevertheless social insects have been spectacularly successful; for example although ants and termites account for only about 2% of known insect species, they form over 50% of the total mass of insects. Their ability to control a territory appears to be the foundation of their success.

The sacrifice of breeding opportunities by most individuals has long been explained as a consequence of these species' unusual haplodiploid method of sex determination, which has the paradoxical consequence that two sterile worker daughters of the same queen share more genes with each other than they would with their offspring if they could breed. However Wilson and Hölldobler argue that this explanation is faulty: for example, it is based on kin selection, but there is no evidence of nepotism in colonies that have multiple queens.

Instead, they write, eusociality evolves only in species that are under strong pressure from predators and competitors, but in environments where it is possible to build "fortresses"; after colonies have established this security, they gain other advantages though co-operative foraging. In support of this explanation they cite the appearance of eusociality in bathyergid mole rats, which are not haplodiploid.

The earliest fossils of insects have been found in Early Devonian rocks from about 400 million years ago, which preserve only a few varieties of flightless insect. The Mazon Creek lagerstätten from the Late Carboniferous, about 300 million years ago, include about 200 species, some gigantic by modern standards, and indicate that insects had occupied their main modern ecological niches as herbivores, detritivores and insectivores. Social termites and ants first appear in the Early Cretaceous, and advanced social bees have been found in Late Cretaceous rocks but did not become abundant until the Mid Cenozoic.

Humans

Modern humans evolved from a lineage of upright-walking apes that has been traced back over 6 million years ago to *Sahelanthropus*. The first known stone tools were made about 2.5 million years ago, apparently by *Australopithecus garhi*, and were found near animal bones that bear scratches made by these tools. The earliest hominines had chimp-sized brains, but there has been a fourfold increase in the last 3 million years; a statistical analysis suggests that hominine brain sizes depend almost completely on the date of the fossils, while the species to which they are assigned has only slight influence.

There is a long-running debate about whether modern humans evolved all over the world simultaneously from existing advanced hominines or are descendants of a single small population in Africa, which then migrated all over the world less than 200,000 years ago and replaced previous hominine species. There is also debate about whether anatomically-modern humans had an intellectual, cultural and technological "Great Leap Forward" under 100,000 years ago and, if so, whether this was due to neurological changes that are not visible in fossils.

Mass Extinctions

Life on Earth has suffered occasional mass extinctions at least since 542 million years ago. Although they are disasters at the time, mass extinctions have sometimes accelerated the evolution of life on Earth. When dominance of particular ecological niches passes from one group of organisms to another, it is rarely because the new dominant group is "superior" to the old and usually because an extinction event eliminates the old dominant group and makes way for the new one. The fossil record appears to show that the gaps between mass extinctions are becoming longer and the average and background rates of extinction are decreasing. Both of these phenomena could be explained in one or more ways:

- The oceans may have become more hospitable to life over the last 500 million years and less vulnerable to mass extinctions: dissolved oxygen became more widespread and penetrated to greater depths; the development of life on land reduced the run-off of nutrients and hence the risk of eutrophication and anoxic events; and marine ecosystems became more diversified so that food chains were less likely to be disrupted.
- Reasonably complete fossils are very rare, most extinct organisms are represented only by partial fossils, and complete fossils are rarest in the oldest rocks. So paleontologists have mistakenly assigned parts of the same organism to different genera which were often defined solely to accommodate these finds – the story of *Anomalocaris* is an example of this. The risk of this mistake is higher for older fossils because these are often unlike parts of any living organism. Many of the "superfluous" genera are represented by fragments which are not found again and the "superfluous" genera appear to become extinct very quickly.

Biodiversity in the fossil record, which is, "the number of distinct genera alive at any given time; that is, those whose first occurrence predates and whose last occurrence postdates that time" shows a different trend: a fairly swift rise from 542 to 400 million years ago; a slight decline from 400 to 200 million years ago, in which the devastating Permian–Triassic extinction event is an important factor; and a swift rise from 200 million years ago to the present.

Abiogenesis

In natural science, abiogenesis or biopoesis is the study of how life arises from inanimate matter through natural processes, and the method by which life on Earth arose. Most amino acids, often called "the building blocks of life", can form via natural chemical reactions unrelated to life, as demonstrated in the Miller–Urey experiment and similar experiments, which involved simulating some of the conditions of the early Earth, in a scientific laboratory. In all living things, these amino acids are organized into proteins, and the construction of these proteins is mediated by nucleic acids. Which of these organic molecules first arose and how they formed the first life is the focus of abiogenesis.

In any theory of abiogenesis, two aspects of life have to be accounted for: replication, and metabolism. The question of which came first gave rise to different types of theories. In the beginning, metabolism-first theories (Oparin coacervate) were proposed, and only later thinking gave rise to the modern, replication-first approach. In modern, still somewhat limited understanding, the first living things on Earth are thought to be single cell prokaryotes (which lack a cell nucleus), perhaps evolved from protobionts (organic molecules surrounded by a membrane-like structure).

The oldest ancient fossil microbe-like objects are dated to be 3.5 Ga (billion years old), approximately one billion years after the formation of the Earth itself. By 2.4 Ga, the ratio of stable isotopes of carbon, iron and sulfur shows the action of living things on inorganic minerals and sediments and molecular biomarkers indicate photosynthesis, demonstrating that life on Earth was widespread by this time. The sequence of chemical events that led to the first nucleic acids is not known. Several hypotheses about early life have been proposed, most notably the iron-sulfur world theory (metabolism without genetics) and the RNA world hypothesis (RNA life-forms).

Conceptual History

Spontaneous Generation

Until the early 19th century, people generally believed in the ongoing spontaneous generation of certain forms of life from non-living matter. This was paired with heterogenesis, the belief that one form of life derives from a different form (*e.g.* bees from flowers). Classical notions of abiogenesis, now more precisely known as *spontaneous generation*, held that certain complex, living organisms are generated by decaying organic substances. According to Aristotle it was a readily observable truth that aphids arise from the dew which falls on plants, flies from putrid matter, mice from dirty hay, crocodiles from rotting logs at the bottom of bodies of water, and so on.

In the 17th century, such assumptions started to be questioned; for example, in 1646, Sir Thomas Browne published his *Pseudodoxia Epidemica* (subtitled *Enquiries into Very many Received Tenets, and Commonly Presumed Truths*), which was an attack on false beliefs and "vulgar errors." His conclusions were not widely accepted. For example, his contemporary, Alexander Ross wrote: "To question this (i.e., spontaneous generation) is to question reason, sense and experience.

If he doubts of this let him go to Egypt, and there he will find the fields swarming with mice, begot of the mud of Nylus, to the great calamity of the inhabitants." In 1665, Robert Hooke published the first drawings of a microorganism. Hooke was followed in 1676 by Anton van Leeuwenhoek, who drew and described microorganisms that are now thought to have been protozoa and bacteria. Many felt the existence of microorganisms was evidence in support of spontaneous generation, since microorganisms seemed too simplistic for sexual reproduction, and asexual reproduction through cell division had not yet been observed.

The first solid evidence against spontaneous generation came in 1668 from Francesco Redi, who proved that no maggots appeared in meat when flies were prevented from laying eggs. It was gradually shown that, at least in the case of all the higher and readily visible organisms, the previous sentiment regarding spontaneous generation was false. The alternative seemed to be biogenesis: that every living thing came from a pre-existing living thing (*omne vivum ex ovo*, Latin for "every living thing from an egg").

In 1768, Lazzaro Spallanzani demonstrated that microbes were present in the air, and could be killed by boiling. In 1861, Louis Pasteur performed a series of experiments which demonstrated that organisms such as bacteria and fungi do not spontaneously appear in sterile, nutrient-rich media.

Pasteur and Darwin

By the middle of the 19th century, the theory of biogenesis had accumulated so much evidential support, due to the work of Louis Pasteur and others, that the alternative theory of spontaneous generation had been effectively disproven. Pasteur himself remarked, after a definitive finding in 1864, "Never will the doctrine of spontaneous generation recover from the mortal blow struck by this simple experiment." The collapse of spontaneous generation, however, left a vacuum of scientific thought on the question of how life *had* first arisen. In a letter to Joseph Dalton Hooker on February 1, 1871, Charles Darwin addressed the question, suggesting that the original spark of life may have begun in a "warm little pond, with all sorts of ammonia and phosphoric salts, lights, heat, electricity, etc. present, so that a protein compound was chemically formed ready to undergo still more complex changes". He went on to explain that "at the present day such matter would be instantly devoured or absorbed, which would not have been the case before living creatures were formed." In other words, the presence of life itself makes the search for the origin of life dependent on the sterile conditions of the laboratory.

"Primordial Soup" Theory

No new notable research or theory on the subject appeared until 1924, when Alexander Oparin reasoned that atmospheric oxygen prevents the synthesis of certain organic compounds that are necessary building blocks for the evolution of life. In his *The Origin of Life*, Oparin proposed that the "spontaneous generation of life" that had been attacked by Louis Pasteur, did in fact occur once, but was now impossible because the conditions found in the early earth had changed, and the presence of living organisms would immediately consume any spontaneously generated organism.

Oparin argued that a "primeval soup" of organic molecules could be created in an oxygen-less atmosphere through the action of sunlight. These would combine in ever-more complex fashions until they formed coacervate droplets. These droplets would "grow" by fusion with other droplets, and "reproduce" through fission into daughter droplets, and so have a primitive metabolism in which those factors which promote "cell integrity" survive, and those that do not become extinct. Many modern theories of the origin of life still take Oparin's ideas as a starting point.

Around the same time, J. B. S. Haldane suggested that the Earth's pre-biotic oceans–very different from their modern counterparts–would have formed a "hot dilute soup" in which organic compounds could have formed. This idea was called *biopoiesis* or *biopoesis*, the process of living matter evolving from self-replicating but nonliving molecules.

Early Conditions

Morse and MacKenzie have suggested that oceans may have appeared first in the Hadean eon, as soon as two hundred million years (200 Ma) after the Earth was formed, in a hot 100 °C (212 °F) reducing environment, and that the pH of about 5.8 rose rapidly towards neutral. This has been supported by Wilde who has pushed the date of the zircon crystals found in the metamorphosed quartzite of Mount Narryer in Western Australia, previously thought to be 4.1–4.2 Ga, to 4.404 Ga. This means that oceans and continental crust existed within 150 Ma of Earth's formation.

Despite this, the Hadean environment was one highly hazardous to life. Frequent collisions with large objects, up to 500 kilometres (310 mi) in diameter, would have been sufficient to vaporise the ocean within a few months of impact, with hot steam mixed with rock vapour leading to high altitude clouds completely covering the planet. After a few months the height of these clouds would have begun to decrease but the cloud base would still have been elevated for about the next thousand years. After that, it would have begun to rain at low altitude. For another two thousand years rains would slowly have drawn down the height of the clouds, returning the oceans to their original depth only 3,000 years after the impact event.

Between 3.8 and 4.1 Ga, changes in the orbits of the gaseous giant planets may have caused a late heavy bombardment that pockmarked the moon and other inner planets (Mercury, Mars, and presumably Earth and Venus). This would likely have sterilized the planet had life appeared before that time.

By examining the time interval between such devastating environmental events, the time interval when life might first have come into existence can be found for different early environments. The study by Maher and Stevenson shows that if the deep marine hydrothermal setting provides a suitable site for the origin of life, abiogenesis could have happened as early as 4.0 to 4.2 Ga, whereas if it occurred at the surface of the earth abiogenesis could only have occurred between 3.7 and 4.0 Ga.

Other research suggests a colder start to life. Work by Leslie Orgel and colleagues on the synthesis of purines has shown that freezing temperatures are advantageous, due to the concentrating effect for key precursors such as hydrogen cyanide. Research by Stanley Miller and colleagues suggested that while adenine and guanine require freezing conditions for synthesis, cytosine and uracil may require boiling temperatures. Based on this research, Miller suggested a beginning of life involving freezing conditions and exploding meteorites. An article in Discover Magazine points to research by the Miller group indicating the formation of seven different amino acids and 11 types of nucleobases in ice when ammonia and cyanide were left in a freezer from 1972–1997. This article also describes research by Christof Biebricher showing the formation of RNA molecules 400 bases long under freezing conditions using an RNA template, a single-strand chain of RNA that guides the formation of a new strand of RNA. As that new RNA strand grows, it adheres to the template.

The explanation given for the unusual speed of these reactions at such a low temperature is eutectic freezing. As an ice crystal forms, it stays pure: only molecules of water join the growing crystal, while impurities like salt or cyanide are excluded. These impurities become crowded in microscopic pockets of liquid within the ice, and this crowding causes the molecules to collide more often.

Evidence of the early appearance of life comes from the Isua supercrustal belt in Western Greenland and from similar formations in the nearby Akilia Islands. Carbon entering into rock formations has a ratio of Carbon-13 (^{13}C) to Carbon-12 (^{12}C) of about "5.5 (in units of $\delta^{13}C$), where because of a preferential biotic uptake of ^{12}C, biomass has a $\delta^{13}C$ of between "20 and "30.

These isotopic fingerprints are preserved in the sediments, and Mojzis has used this technique to suggest that life existed on the planet already by 3.85 billion years ago. Lazcano and Miller (1994) suggest that the rapidity of the evolution of life is dictated by the rate of recirculating water through mid-ocean submarine vents.

Complete recirculation takes 10 million years, thus any organic compounds produced by then would be altered or destroyed by temperatures exceeding 300 °C (572 °F). They estimate that the development of a 100 kilobase genome of a DNA/protein primitive heterotroph into a 7000 gene filamentous cyanobacterium would have required only 7 Ma.

Current Models

There is no truly "standard model" of the origin of life. Most currently accepted models draw at least some elements from the framework laid out by the Oparin-Haldane hypothesis. Under that umbrella, however, are a wide array of disparate discoveries and conjectures such as the following, listed in a rough order of postulated emergence:

1. Some theorists suggest that the atmosphere of the early Earth may have been chemically reducing in nature, composed primarily of methane (CH_4), ammonia (NH_3), water (H_2O),

hydrogen sulfide (H_2S), carbon dioxide (CO_2) or carbon monoxide (CO), and phosphate (PO_4^{3-}), with molecular oxygen (O_2) and ozone (O_3) either rare or absent.

2. In such a reducing atmosphere, electrical activity can catalyse the creation of certain basic small molecules (monomers) of life, such as amino acids. This was demonstrated in the Miller–Urey experiment by Stanley L. Miller and Harold C. Urey in 1953.
3. Phospholipids (of an appropriate length) can spontaneously form lipid bilayers, a basic component of the cell membrane.
4. A fundamental question is about the nature of the first self-replicating molecule. Since replication is accomplished in modern cells through the cooperative action of proteins and nucleic acids, the major schools of thought about how the process originated can be broadly classified as "proteins first" and "nucleic acids first".
5. The principal thrust of the "nucleic acids first" argument is as follows:
    1. The polymerization of nucleotides into random RNA molecules might have resulted in self-replicating ribozymes (RNA world hypothesis)
    2. Selection pressures for catalytic efficiency and diversity might have resulted in ribozymes which catalyse peptidyl transfer (hence formation of small proteins), since oligopeptides complex with RNA to form better catalysts. The first ribosome might have been created by such a process, resulting in more prevalent protein synthesis.
    3. Synthesized proteins might then outcompete ribozymes in catalytic ability, and therefore become the dominant biopolymer, relegating nucleic acids to their modern use, predominantly as a carrier of genomic information.

As of 2010, no one has yet synthesized a "protocell" using basic components which would have the necessary properties of life (the so-called *"bottom-up-approach"*). Without such a proof-of-principle, explanations have tended to be short on specifics. However, some researchers are working in this field, notably Steen Rasmussen at Los Alamos National Laboratory and Jack Szostak at Harvard University. Others have argued that a *"top-down approach"* is more feasible. One such approach, successfully attempted by Craig Venter and others at The Institute for Genomic Research, involves engineering existing prokaryotic cells with progressively fewer genes, attempting to discern at which point the most minimal requirements for life were reached. The biologist John Desmond Bernal coined the term Biopoesis for this process, and suggested that there were a number of clearly defined "stages" that could be recognised in explaining the origin of life.

- Stage 1: The origin of biological monomers
- Stage 2: The origin of biological polymers
- Stage 3: The evolution from molecules to cell.

Bernal suggested that evolution may have commenced early, some time between Stage 1 and 2.

Regarding Monomer Formation

One of the most important pieces of experimental support for the "soup" theory came in 1953. A graduate student, Stanley Miller, and his professor, Harold Urey, performed an experiment that demonstrated how organic molecules could have spontaneously formed from inorganic precursors, under conditions like those posited by the Oparin-Haldane Hypothesis. The now-famous "Miller–Urey experiment" used a highly reduced mixture of gases–methane, ammonia and hydrogen–to form basic organic monomers, such as amino acids. This provided direct experimental support for the second point of the "soup" theory, and it is around the remaining two points of the theory that much of the debate now centres.

Apart from the Miller–Urey experiment, the next most important step in research on prebiotic organic synthesis was the demonstration by Joan Oró that the nucleic acid purine base, adenine, was formed by heating aqueous ammonium cyanide solutions. In support of abiogenesis in eutectic ice, more recent work demonstrated the formation of s-triazines (alternative nucleobases), pyrimidines (including cytosine and uracil), and adenine from urea solutions subjected to freeze-thaw cycles under a reductive atmosphere (with spark discharges as an energy source).

Regarding Monomer Accumulation

The "soup" theory relies on the assumption proposed by Darwin that in an environment with no pre-existing life, organic molecules may have accumulated and provided an environment for chemical evolution.

Regarding Further Transformation

The spontaneous formation of complex polymers from abiotically generated monomers under the conditions posited by the "soup" theory is not at all a straightforward process. Besides the necessary basic organic monomers, compounds that would have prohibited the formation of polymers were formed in high concentration during the Miller–Urey and Oró experiments. The Miller experiment, for example, produces many substances that would undergo cross-reactions with the amino acids or terminate the peptide chain.

More fundamentally, it can be argued that the most crucial challenge unanswered by this theory is how the relatively simple organic building blocks polymerise and form more complex structures, interacting in consistent ways to form a protocell. For example, in an aqueous environment hydrolysis of oligomers/polymers into their constituent monomers would be favoured over the condensation of individual monomers into polymers.

The Deep Sea vent Theory

The deep sea vent, or hydrothermal vent, theory for the origin of life on Earth posits that life may have begun at submarine hydrothermal vents, where hydrogen-rich fluids emerge from below the sea floor and interface with carbon dioxide-rich ocean water. Sustained chemical energy in such systems is derived from redox reactions, in which electron donors, such as molecular hydrogen, react with electron acceptors, such as carbon dioxide.

Fox's Experiments

In the 1950s and 1960s, Sidney W. Fox studied the spontaneous formation of peptide structures under conditions that might plausibly have existed early in Earth's history. He demonstrated that amino acids could spontaneously form small peptides. These amino acids and small peptides could be encouraged to form closed spherical membranes, called protenoid microspheres, which show many of the basic characteristics of 'life'.

Eigen's Hypothesis

In the early 1970s the problem of the origin of life was approached by Manfred Eigen and Peter Schuster of the Max Planck Institute for Biophysical Chemistry. They examined the transient stages between the molecular chaos and a self-replicating hypercycle in a prebiotic soup. In a hypercycle, the information storing system (possibly RNA) produces an enzyme, which catalyzes the formation of another information system, in sequence until the product of the last aids in the formation of the first information system. Mathematically treated, hypercycles could create quasispecies, which through natural selection entered into a form of Darwinian evolution. A boost to hypercycle theory was the discovery that RNA, in certain circumstances, forms itself into ribozymes, capable of catalysing their own chemical reactions.

However, these reactions are limited to self-excisions (in which a longer RNA molecule becomes shorter), and much rarer small additions that are incapable of coding for any useful protein. The hypercycle theory is further degraded since the hypothetical RNA would require the existence of complex biochemicals such as nucleotides which are not formed under the conditions proposed by the Miller–Urey experiment.

Hoffmann's Contributions

Geoffrey W. Hoffmann, a student of Eigen, contributed to the concept of life involving both replication and metabolism emerging from catalytic noise. His contributions included showing that an early sloppy translation machinery can be stable against an error catastrophe of the type that had been envisaged as problematical by Leslie Orgel ("Orgel's paradox") and calculations regarding the occurrence of a set of required catalytic activities together with the exclusion of catalytic activities that would be disruptive. This is called the stochastic theory of the origin of life.

Wächtershäuser's Hypothesis

Another possible answer to this polymerization conundrum was provided in 1980s by the German chemist Günter Wächtershäuser, in his iron-sulfur world theory. In this theory, he postulated the evolution of (bio)chemical pathways as fundamentals of the evolution of life. Moreover, he presented a consistent system of tracing today's biochemistry back to ancestral reactions that provide alternative pathways to the synthesis of organic building blocks from simple gaseous compounds.

In contrast to the classical Miller experiments, which depend on external sources of energy (such as simulated lightning or UV irradiation), "Wächtershäuser systems" come with a built-in source of energy, sulfides of iron and other minerals (e.g. pyrite). The energy released from redox reactions of these metal sulfides is not only available for the synthesis of organic molecules, but also for the formation of oligomers and polymers. It is therefore hypothesized that such systems may be able to evolve into autocatalytic sets of self-replicating, metabolically active entities that would predate the life forms known today.

The experiment produced a relatively small yield of dipeptides (0.4% to 12.4%) and a smaller yield of tripeptides (0.10%) but the authors also noted that: "under these same conditions dipeptides hydrolysed rapidly."

Radioactive Beach Hypothesis

Zachary Adam at the University of Washington, Seattle, claims that stronger tidal processes from a much closer moon may have concentrated grains of uranium and other radioactive elements at the high water mark on primordial beaches where they may have been responsible for generating life's building blocks. According to computer models reported in *Astrobiology*, a deposit of such radioactive materials could show the same self-sustaining nuclear reaction as that found in the Oklo uranium ore seam in Gabon. Such radioactive beach sand provides sufficient energy to generate organic molecules, such as amino acids and sugars from acetonitrile in water.

Radioactive monazite also releases soluble phosphate into regions between sand-grains, making it biologically "accessible". Thus amino acids, sugars and soluble phosphates can all be simultaneously produced, according to Adam. Radioactive actinides, then in greater concentrations, could have formed part of organo-metallic complexes.

These complexes could have been important early catalysts to living processes. John Parnell of the University of Aberdeen suggests that such a process could provide part of the "crucible of life" on any early wet rocky planet, so long as the planet is large enough to have generated a system of plate tectonics which brings radioactive minerals to the surface. As the early Earth is believed to have had many smaller "platelets" it would provide a suitable environment for such processes.

Thermodynamic Origin of Life: Ultraviolet and Temperature-Assisted Replication (UVTAR) Model

Karo Michaelian of the National Autonomous University of Mexico (UNAM) points out that any model for the origin of life must take into account the fact that life is an irreversible thermodynamic process which arises and persists to produce entropy. Entropy production is not incidental to the process of life, but rather the fundamental reason for its existence. Present day life augments the entropy production of Earth by catalysing the water cycle through evapotranspiration. Michaelian argues that if the thermodynamic function of life today is to produce entropy through coupling with the water cycle, then this probably was its function at its very beginnings. It turns out that both RNA and DNA when in water solution are very strong absorbers and extremely rapid dissipaters of ultraviolet light within the 200 nm - 300 nm wavelength range, just that high energy part of the sun's spectrum that could have penetrated the dense prebiotic atmosphere.

Cnossen et al. have shown that the amount of UV light reaching the Earth's surface in the Archean could have been up to 31 orders of magnitude larger than it is today at 260 nm where RNA and DNA absorb most strongly.

Absorption and dissipation of UV light by these organic molecules at the Archean ocean surface would have increased significantly the temperature of the surface skin layer leading to enhanced evaporation and thus augmenting the primitive water cycle. Since absorption and dissipation of high energy photons is an entropy producing process, Michaelian argues that non-equilbrium abiogenic synthesis of RNA and DNA utilizing UV light would have been thermodynamically favoured.

A simple mechanism to explain the replication of RNA and DNA without the use of enzymes can also be given within the same thermodynamic framework by assuming that life arose when the temperature of the primitive seas had cooled to somewhat below the denaturing temperature of RNA or DNA (based on the ratio of $^{18}O/^{16}O$ found in cherts of the Barberton greenstone belt of South Africa of about 3.5 to 3.2 Ga., surface temperatures are predicted to have been around 70±15 °C, similar to RNA or DNA denaturing temperatures).

During the night, the surface water temperature would be below the denaturing temperature and single strand RNA/DNA could act as a template for the formation of double strand RNA/DNA. During the daylight hours, RNA and DNA would absorb UV light and convert this directly to heating of the ocean surface, raising the local temperature enough to allow for denaturing of RNA and DNA.

The copying process would be repeated during the cool period overnight. Such a temperature assisted mechanism of replication bears similarity to Polymerase Chain Reaction (PCR), a routine laboratory procedure to multiply DNA segments. Michaelian suggests that traditional origin of life research, expecting to describe the emergence of life from near-equilibrium conditions, is erroneous and that non-equilibrium conditions must be considered, in particular, the importance of entropy production to the emergence of life.

Since denaturation would be most probable in the late afternoon when the Archean sea surface temperature would be highest, and since late afternoon submarine sunlight is somewhat circularly polarized, the homochirality of the organic molecules of life can also be explained within the proposed thermodynamic framework.

Models to Explain Homochirality

Some process in chemical evolution must account for the origin of homochirality, i.e. all building blocks in living organisms having the same "handedness" (amino acids being left-handed, nucleic acid sugars (ribose and deoxyribose) being right-handed, and chiral phosphoglycerides). Chiral molecules can be synthesized, but in the absence of a chiral source or a chiral catalyst, they are formed in a 50/50 mixture of both enantiomers.

This is called a racemic mixture. Clark has suggested that homochirality may have started in space, as the studies of the amino acids on the Murchison meteorite showed L-alanine to be more than twice as frequent as its D form, and L-glutamic acid was more than 3 times prevalent than its D counterpart. It is suggested that polarised light has the power to destroy one enantiomer within the proto-planetary disk. Noyes showed that beta decay caused the breakdown of D-leucine, in a racemic mixture, and that the presence of ^{14}C, present in larger amounts in organic chemicals in the early Earth environment, could have been the cause.

Robert M. Hazen reports upon experiments conducted in which various chiral crystal surfaces act as sites for possible concentration and assembly of chiral monomer units into macromolecules. Once established, chirality would be selected for. Work with organic compounds found on meteorites tends to suggest that chirality is a characteristic of abiogenic synthesis, as amino acids show a left-handed bias, whereas sugars show a predominantly right-handed bias.

Self-organization and Replication

While features of self-organization and self-replication are often considered the hallmark of living systems, there are many instances of abiotic molecules exhibiting such characteristics under proper conditions. For example Martin and Russel show that physical compartmentation by cell membranes from the environment and self-organization of self-contained redox reactions are the most conserved attributes of living things, and they argue therefore that inorganic matter with such attributes would be life's most likely last common ancestor. Virus self-assembly within host cells has implications for the study of the origin of life, as it lends further credence to the hypothesis that life could have started as self-assembling organic molecules.

From Organic Molecules to Protocells

The question "How do simple organic molecules form a protocell?" is largely unanswered but there are many hypotheses. Some of these postulate the early appearance of nucleic acids ("genes-first") whereas others postulate the evolution of biochemical reactions and pathways first ("metabolism-first"). Recently, trends are emerging to create hybrid models that combine aspects of both.

"Genes First" Models: The RNA World

The RNA world hypothesis describes an early Earth with self-replicating and catalytic RNA but no DNA or proteins. This has spurred scientists to try to determine if relatively short RNA molecules could have spontaneously formed that were capable of catalysing their own continuing replication. A number of hypotheses of modes of formation have been put forward.

Early cell membranes could have formed spontaneously from proteinoids, protein-like molecules that are produced when amino acid solutions are heated–when present at the correct concentration in aqueous solution, these form microspheres which are observed to behave similarly to membrane-enclosed compartments. Other possibilities include systems of chemical reactions taking place within clay substrates or on the surface of pyrite rocks.

Factors supportive of an important role for RNA in early life include its ability to act both to store information and catalyse chemical reactions (as a ribozyme); its many important roles as an intermediate in the expression and maintenance of the genetic information (in the form of DNA) in modern organisms; and the ease of chemical synthesis of at least the components of the molecule under conditions approximating the early Earth.

Relatively short RNA molecules which can duplicate others have been artificially produced in the lab. Such replicase RNA, which functions as both code and catalyst provides a template upon which copying can occur. Jack Szostak has shown that certain catalytic RNAs can, indeed, join smaller RNA sequences together, creating the potential, in the right conditions for self-replication. If these were present, Darwinian selection would favour the proliferation of such self-catalysing structures, to which further functionalities could be added. Lincoln and Joyce identified an RNA enzyme capable of self sustained replication. Researchers have pointed out difficulties for the abiotic synthesis of nucleotides from cytosine and uracil. Cytosine has a half-life of 19 days at 100 °C (212 °F) and 17,000 years in freezing water. Larralde et al., say that "the generally accepted prebiotic synthesis of ribose, the formose reaction, yields numerous sugars without any selectivity." and they conclude that their "results suggest that the backbone of the first genetic material could not have contained ribose or other sugars because of their instability." The ester linkage of ribose and phosphoric acid in RNA is known to be prone to hydrolysis.

A slightly different version of the RNA-world hypothesis is that a different type of nucleic acid, such as PNA, TNA or GNA, was the first one to emerge as a self-reproducing molecule, to be replaced by RNA only later. Pyrimidine ribonucleosides and their respective nucleotides have been prebiotically synthesised by a sequence of reactions which by-pass the free sugars, and are assembled in a stepwise fashion by going against the dogma that nitrogenous and oxygenous chemistries should be avoided.

In a series of publications, The Sutherland Group at the School of Chemistry, University of Manchester have demonstrated high yielding routes to cytidine and uridine ribonucleotides built from small 2 and 3 carbon fragments such as glycolaldehyde, glyceraldehyde or glyceraldehyde-3-phosphate, cyanamide and cyanoacetylene. One of the steps in this sequence allows the isolation of enantiopure ribose aminooxazoline if the enantiomeric excess of glyceraldehyde is 60 % or greater.

This can be viewed as a prebiotic purification step, where the said compound spontaneously crystallised out from a mixture of the other pentose aminooxazolines. Ribose aminooxazoline can then react with cyanoacetylene in a mild and highly efficient manner to give the alpha cytidine ribonucleotide. Photoanomerization with UV light allows for inversion about the 1' anomeric centre to give the correct beta stereochemistry.

In 2009 they showed that the same simple building blocks allow access, via phosphate controlled nucleobase elaboration, to 2',3'-cyclic pyrimidine nucleotides directly, which are known to be able to polymerise into RNA. This paper also highlights the possibility for the photo-sanitization of the pyrimidine-2',3'-cyclic phosphates. James Ferris's studies have shown that clay minerals of montmorillonite will catalyse the formation of RNA in aqueous solution, by joining activated mono RNA nucleotides to join together to form longer chains. Although these chains have random sequences, the possibility that one sequence began to non-randomly increase its frequency by increasing the speed of its catalysis is possible to "kick start" biochemical evolution.

Iron-sulfur World

One of the earliest incarnations of this idea was put forward in 1924 with Alexander Oparin's notion of primitive self-replicating vesicles which predated the discovery of the structure of DNA. More recent variants in the 1980s and 1990s include Günter Wächtershäuser's iron-sulfur world theory and models introduced by Christian de Duve based on the chemistry of thioesters. More abstract and theoretical arguments for the plausibility of the emergence of metabolism without the presence of genes include a mathematical model introduced by Freeman Dyson in the early 1980s and Stuart Kauffman's notion of collectively autocatalytic sets.

However, the idea that a closed metabolic cycle, such as the reductive citric acid cycle, could form spontaneously (proposed by Günter Wächtershäuser) remains debated. In an article entitled "Self-Organizing Biochemical Cycles", the late Leslie Orgel summarized his analysis of the proposal by stating, "There is at present no reason to expect that multistep cycles such as the reductive citric acid cycle will self-organize on the surface of FeS/FeS2 or some other mineral."

It is possible that another type of metabolic pathway was used at the beginning of life. For example, instead of the reductive citric acid cycle, the "open" acetyl-CoA pathway (another one of the five recognised ways of carbon dioxide fixation in nature today) would be compatible with the idea of self-organisation on a metal sulfide surface. The key enzyme of this pathway, carbon monoxide dehydrogenase/ acetyl-CoA synthase harbours mixed nickel-iron-sulfur clusters in its reaction centres and catalyses the formation of acetyl-CoA (which may be regarded as a modern form of acetyl-thiol) in a single step.

Thermosynthesis World

Today's bioenergetic process of fermentation is related to the just mentioned citric acid cycle or the Acetyl-CoA pathway that have been connected to the primordial iron-sulfur world. In a different approach, today's bioenergetic process of chemiosmosis, which plays an essential role in cellular respiration and photosynthesis, is considered as more fundamental than fermentation: in Anthonie Muller's "thermosynthesis world" the ATP Synthase enzyme that sustains chemiosmosis is proposed as today's enzyme that is the closest connected to the first metabolic process. First life needed an energy source to bring about the condensation reaction that yielded the peptide bonds of proteins and the phosphodiester bonds of RNA. In a generalization and thermal variation of the binding change mechanism of today's ATP Synthase, the "First Protein" would have bound substrates (peptides, phosphate, nucleosides, RNA 'monomers') and condensed them to a reaction product that remained bound until it after a temperature change was released upon a thermal unfolding.

The energy source of the thermosynthesis world was thermal cycling, the result of suspension of the protocell in a convection current, as is plausible in a volcanic hot spring; the convection accounts for the self-organization and dissipative structure required in any origin of life model. The still ubiquitous role of thermal cycling in germination and cell division is considered a relic of primordial thermosynthesis.

By phosphorylating cell membrane lipids, this 'First Protein' gave a selective advantage to the lipid protocell that contained the protein. In the beginning this First Protein also synthesized a library with many proteins, of which only a minute fraction had thermosynthesis capabilities. Just as proposed by Dyson for the first proteins, the First Protein propagated functionally: it made daughters with similar capabilities, but it did not copy itself. Functioning daughters consisted of different amino acid sequences.

Over a long time, RNA sequences were selected among the at first randomly synthesized RNAs by the criterion of speed and efficiency increase of First Protein synthesis, for instance by the creation of RNA that functioned as messenger RNA, Transfer RNA and ribosomal RNA, or, even more generally, all the components of the RNA World were also generated and selected. The thermosynthesis world therefore in theory accounts for the origin of the genetic machinery.

Whereas the iron-sulfur world identifies a circular pathway as the most simple—and therefore assumes the existence of enzymes—the thermosynthesis world does not even invoke a pathway, and does not assume the existence of regular enzymes: ATP Synthase's binding change mechanism resembles a physical adsorption process that yields free energy, rather than a regular enzyme's mechanism, which decreases the free energy. The RNA World also implies the existence of several enzymes. But even the emergence of a single enzyme by chance is implausible. The thermosynthesis world is therefore more simple, and thus more plausible, than the iron-sulfur and RNA worlds.

Possible Role of Bubbles

Waves breaking on the shore create a delicate foam composed of bubbles. Winds sweeping across the ocean have a tendency to drive things to shore, much like driftwood collecting on the beach. It is possible that organic molecules were concentrated on the shorelines in much the same way. Shallow coastal waters also tend to be warmer, further concentrating the molecules through evaporation. While bubbles composed mostly of water burst quickly, water containing amphiphiles forms much more stable bubbles, lending more time to the particular bubble to perform these crucial reactions.

Amphiphiles are oily compounds containing a hydrophilic head on one or both ends of a hydrophobic molecule. Some amphiphiles have the tendency to spontaneously form membranes in water. A spherically closed membrane contains water and is a hypothetical precursor to the modern cell membrane. If a protein would increase the integrity of its parent bubble, that bubble had an advantage, and was placed at the top of the natural selection waiting list. Primitive reproduction can be envisioned when the bubbles burst, releasing the results of the 'experiment' into the surrounding medium. Once enough of the 'right stuff' was released into the medium, the development of the first prokaryotes, eukaryotes, and multicellular organisms could be achieved.

Similarly, bubbles formed entirely out of protein-like molecules, called microspheres, will form spontaneously under the right conditions. But they are not a likely precursor to the modern cell membrane, as cell membranes are composed primarily of lipid compounds rather than amino-acid compounds (for types of membrane spheres associated with abiogenesis. A recent model by Fernando and Rowe suggests that the enclosure of an autocatalytic non-enzymatic metabolism within protocells may have been one way of avoiding the side-reaction problem that is typical of metabolism first models.

Other Models

Autocatalysis

In 1993 Stuart Kauffman proposed that life initially arose as autocatalytic chemical networks. British ethologist Richard Dawkins wrote about autocatalysis as a potential explanation for the origin of life in his 2004 book *The Ancestor's Tale*. Autocatalysts are substances which catalyse the production of themselves, and therefore have the property of being a simple molecular replicator. In his book, Dawkins cites experiments performed by Julius Rebek and his colleagues at the Scripps Research Institute in California in which they combined amino adenosine and pentafluorophenyl ester with the autocatalyst amino adenosine triacid ester (AATE). One system from the experiment contained variants of AATE which catalysed the synthesis of themselves. This experiment demonstrated the possibility that autocatalysts could exhibit competition within a population of entities with heredity, which could be interpreted as a rudimentary form of natural selection.

Clay Theory

A model for the origin of life based on clay was forwarded by A. Graham Cairns-Smith of the University of Glasgow in 1985 and explored as a plausible illustration by several other scientists, including Richard Dawkins. Clay theory postulates that complex organic molecules arose gradually on a pre-existing, non-organic replication platform—silicate crystals in solution. Complexity in companion molecules developed as a function of selection pressures on types of clay crystal is then exapted to serve the replication of organic molecules independently of their silicate "launch stage". Cairns-Smith is a staunch critic of other models of chemical evolution. However, he admits that like many models of the origin of life, his own also has its shortcomings (Horgan 1991).

In 2007, Kahr and colleagues reported their experiments to examine the idea that crystals can act as a source of transferable information, using crystals of potassium hydrogen phthalate. "Mother" crystals with imperfections were cleaved and used as seeds to grow "daughter" crystals from solution. They then examined the distribution of imperfections in the crystal system and found that the imperfections in the mother crystals were indeed reproduced in the daughters. The daughter crystals had many additional imperfections. For a gene-like behaviour the additional imperfections should be much less than the parent ones, thus Kahr concludes that the crystals "were not faithful enough to store and transfer information from one generation to the next".

Gold's "Deep-hot Biosphere" Model

In the 1970s, Thomas Gold proposed the theory that life first developed not on the surface of the Earth, but several kilometers below the surface. The discovery in the late 1990s of nanobes (filamental structures that are smaller than bacteria, but that may contain DNA) in deep rocks might be seen as lending support to Gold's theory. It is now reasonably well established that microbial life is plentiful at shallow depths in the Earth, up to 5 kilometres (3.1 mi) below the surface, in the form of extremophile archaea, rather than the better-known eubacteria (which live in more accessible conditions).

It is claimed that discovery of microbial life below the surface of another body in our solar system would lend significant credence to this theory. Thomas Gold also asserted that a trickle of food from a deep, unreachable, source is needed for survival because life arising in a puddle of organic material is likely to consume all of its food and become extinct. Gold's theory is that flow of food is due to out-gassing of primordial methane from the Earth's mantle; more conventional explanations of the food supply of deep microbes (away from sedimentary carbon compounds) is that the organisms subsist on hydrogen released by an interaction between water and (reduced) iron compounds in rocks.

"Primitive" Extraterrestrial Life

An alternative to Earthly abiogenesis is the hypothesis that primitive life may have originally formed extraterrestrially, either in space or on a nearby planet (Mars). (Note that exogenesis is related to, but not the same as, the notion of panspermia). A supporter of this theory was Francis Crick.

Organic compounds are relatively common in space, especially in the outer solar system where volatiles are not evaporated by solar heating. Comets are encrusted by outer layers of dark material, thought to be a tar-like substance composed of complex organic material formed from simple carbon compounds after reactions initiated mostly by irradiation by ultraviolet light. It is supposed that a rain of material from comets could have brought significant quantities of such complex organic molecules to Earth.

An alternative but related hypothesis, proposed to explain the presence of life on Earth so soon after the planet had cooled down, with apparently very little time for prebiotic evolution, is that life formed first on early Mars. Due to its smaller size Mars cooled before Earth (a difference of hundreds of millions of years), allowing prebiotic processes there while Earth was still too hot. Life was then transported to the cooled Earth when crustal material was blasted off Mars by asteroid and comet impacts. Mars continued to cool faster and eventually became hostile to the continued evolution or even existence of life (it lost its atmosphere due to low volcanism); Earth is following the same fate as Mars, but at a slower rate.

Neither hypothesis actually answers the question of how life first originated, but merely shifts it to another planet or a comet. However, the advantage of an extraterrestrial origin of primitive life is that life is not required to have evolved on each planet it occurs on, but rather in a single location, and then spread about the galaxy to other star systems via cometary and/or meteorite impact. Evidence to support the hypothesis is scant, but it finds support in recent study of Martian meteorites found in Antarctica and in studies of extremophile microbes. Additional support comes from a recent discovery of a bacterial ecosystem whose energy source is radioactivity.

A 2001 experiment led by Jason Dworkin subjected a frozen mixture of water, methanol, ammonia and carbon monoxide to UV radiation, mimicking conditions found in an extraterrestrial environment. This combination yielded large amounts of organic material that self-organised to form bubbles or micelles when immersed in water. Dworkin considered these bubbles to resemble cell membranes that enclose and concentrate the chemistry of life, separating their interior from the outside world.

The bubbles produced in these experiments were between 10 to 40 micrometres (0.00039 to 0.0016 in), or about the size of red blood cells. Remarkably, the bubbles fluoresced, or glowed, when exposed to UV light. Absorbing UV and converting it into visible light in this way was considered one possible way of providing energy to a primitive cell. If such bubbles played a role in the origin of life, the fluorescence could have been a precursor to primitive photosynthesis. Such fluorescence also provides the benefit of acting as a sunscreen, diffusing any damage that otherwise would be inflicted by UV radiation. Such a protective function would have been vital for life on the early Earth, since the ozone layer, which blocks out the sun's most destructive UV rays, did not form until after photosynthetic life began to produce oxygen.

Extraterrestrial Amino Acids

Another idea is that amino acids which were formed extra-terrestrially arrived on Earth via comets. In 2009 it was announced by NASA that scientists have identified one of the fundamental chemical buildings blocks of life in a comet for the first time: glycine, an amino acid, was detected in the material ejected from Comet Wild-2 in 2004 and grabbed by NASA's Stardust probe. Tiny grains, just a few thousandths of a millimetre in size, were collected from the comet and returned to Earth in 2006 in a sealed capsule, and distributed among the world's leading astro-biology labs. NASA said in a statement that it took some time for the investigating team, led by Dr Jamie Elsila, to convince itself that the glycine signature found in Stardust's sample bay was genuine and not just Earthly contamination.

Glycine has been detected in meteorites before and there are also observations in interstellar gas clouds claimed for telescopes, but the Stardust find is described as a first in cometary material. It is known that prior to the emergence of life on Earth, the early solar system's planets were regularly bombarded by comets. Dr. Carl Pilcher, who leads NASA's Astrobiology Institute commented that "The discovery of glycine in a comet supports the idea that the fundamental building blocks of life are prevalent in space, and strengthens the argument that life in the Universe may be common rather than rare."

Lipid World

This theory postulates that the first self-replicating object was lipid-like. It is known that phospholipids form bilayers in water while under agitation– the same structure as in cell membranes. These molecules were not present on early Earth, however other amphiphilic long chain molecules also form membranes. Furthermore, these bodies may expand (by insertion of additional lipids), and under excessive expansion may undergo spontaneous splitting which preserves the same size and composition of lipids in the two progenies. The main idea in this theory is that the molecular composition of the lipid bodies is the preliminary way for information storage, and evolution led to the appearance of polymer entities such as RNA or DNA that may store information favorably. Still, no biochemical mechanism has been offered to support the Lipid World theory.

Polyphosphates

The problem with most scenarios of abiogenesis is that the thermodynamic equilibrium of amino acid versus peptides is in the direction of separate amino acids. What has been missing is some force that drives polymerization. The resolution of this problem may well be in the properties of polyphosphates. Polyphosphates are formed by polymerization of ordinary monophosphate ions PO_4^{-3}.

Several mechanisms for such polymerization have been suggested. Polyphosphates cause polymerization of amino acids into peptides. They are also logical precursors in the synthesis of such key biochemical compounds as ATP. A key issue seems to be that calcium reacts with soluble phosphate to form insoluble calcium phosphate (apatite), so some plausible mechanism must be found to keep calcium ions from causing precipitation of phosphate. There has been much work on this topic over the years, but an interesting new idea is that meteorites may have introduced reactive phosphorus species on the early Earth.

PAH World Hypothesis

Other sources of complex molecules have been postulated, including extraterrestrial stellar or interstellar origin. For example, from spectral analyses, organic molecules are known to be present in comets and meteorites. In 2004, a team detected traces of polycyclic aromatic hydrocarbons (PAH's) in a nebula. Those are the most complex molecules so far found in space. The use of PAH's has also been proposed as a precursor to the RNA world in the PAH world hypothesis.

The Spitzer Space Telescope has recently detected a star, HH 46-IR, which is forming by a process similar to that by which the sun formed. In the disk of material surrounding the star, there is a very large range of molecules, including cyanide compounds, hydrocarbons, and carbon monoxide. PAHs have also been found all over the surface of galaxy M81, which is 12 million light years away from the Earth, confirming their widespread distribution in space.

Multiple Genesis

Different forms of life may have appeared quasi-simultaneously in the early history of Earth. The other forms may be extinct, leaving distinctive fossils through their different biochemistry (e.g., using arsenic instead of phosphorus), survive as extremophiles, or simply be unnoticed through their being analogous to organisms of the current life tree. Hartman for example combines a number of theories together, by proposing that:

The first organisms were self-replicating iron-rich clays which fixed carbon dioxide into oxalic and other dicarboxylic acids. This system of replicating clays and their metabolic phenotype then evolved into the sulfide rich region of the hotspring acquiring the ability to fix nitrogen. Finally phosphate was incorporated into the evolving system which allowed the synthesis of nucleotides and phospholipids. If biosynthesis recapitulates biopoesis, then the synthesis of amino acids preceded the synthesis of the purine and pyrimidine bases.

Furthermore the polymerization of the amino acid thioesters into polypeptides preceded the directed polymerization of amino acid esters by polynucleotides. Lynn Margulis's endosymbiotic theory suggests that multiple forms of bacteria entered into symbiotic relationship to form the eukaryotic cell. The horizontal transfer of genetic material between bacteria promotes such symbiotic relationships, and thus many separate organisms may have contributed to building what has been recognised as the Last Universal Common Ancestor (LUCA) of modern organisms. James Lovelock's Gaia theory, proposes that such bacterial symbiosis establishes the environment as a system produced by and supportive of life. His arguments strongly weaken the case for life having evolved elsewhere in the solar system.

Life may seem biologically, socially, and culturally complex, but physically and chemically it's rather simple. When reduced to its component parts, the basic ingredients of life—any life, from bacteria to whales, and including humans too—are hardly more exotic than two dozen molecules. So to understand the essential properties of life, we need not examine an organism as messy as the entire human body; the molecular nature of contemporary life will do. This is the realm of biochemistry—the study of chemical processes in living organisms. All living systems are made of cells, the simplest form of material substance having the common attributes of life—birth, metabolism, and death. From primitive microbes to intelligent humans, the basic unit is the cell. To appreciate chemical evolution—the changes that occurred among atoms and molecules in order to produce life in this, the fifth, CHEMICAL EPOCH—we need only consider the construction of a cell.

Cells are minute, ~100 times smaller than a millimeter (or 10 microns across), thus invisible to the naked eye. A microscopic view shows a central nucleus to be the most complex part of most (but not all) cells. Containing trillions of atoms and molecules, such biological nuclei should not be confused with the much smaller atomic nuclei produced in the cores of stars. Resembling the yolk of an egg, a cell's biological nucleus is surrounded by a thick, fluidic cytoplasm of less complexity. The whole unicell is encased within a semi-permeable membrane through which atoms and molecules can pass in and out.

Cells, then, are the simplest form of life—the "bricks" of anatomical structure. However, they're vastly more complex than the simplest form of matter—elementary particles within atoms. In fact, it's worth stressing that simple cells are more complex than any known type of inanimate matter, lending credence to the evolutionary progression from simplicity to complexity, from matter to life, along the arrow of time. One of the primary creatures of all, the amoeba, consists of only one cell. More advanced organisms usually contain many additional cells, often huge clusters of them. A grown human, for example, harbors ~10^{14} (*i.e.*, a hundred trillion) microscopic cells in the guts, skin, bones, hair, muscles, and every other part of our bodies (though only a tenth, or ten trillion, of these are true human cells, the other 90% being bacterial cells that crawl around in our bodies). Each one of these cells furthermore contains very large numbers—trillions or more—of atoms and molecules. The density of basic matter in advanced life forms is indeed great—~200 million cells/cm^3 —which is why some researchers regard cellular compactness as a rudimentary measure of complexity. But surely there's more to the idea of complexity than just structural density.

Over the course of time, even as brief as a second, large numbers of cells are destroyed owing to the normal process of aging and death. All living systems are nonetheless able to maintain a reasonably constant size and appearance throughout adulthood. Thus, while some cells are dying, others must be forming. Our bodies and those of all other living creatures continually manufacture cell nuclei, cytoplasm, and membrane to sustain themselves throughout life. They do so by means of a curious interaction between the two basic building blocks of life. Amino Acids The dominant, foundational ingredient of the cytoplasm in any living system is protein, a word deriving from the Greek and meaning "of first importance." Not the name of a particular substance, rather the term for an entire class of molecules, proteins contain large quantities of the element carbon. In fact, 50% of the dry weight of our bodies is carbon, largely because each of us harbors tens of thousands of proteins.

Such animate, "organic" substances strongly contrast with things obviously not living, such as a slab of concrete or a pinch of salt. Those things are said to be "inorganic," for they're made mostly of minerals. Inanimate objects have no proteins and their carbon content often amounts to less than ~0.1% of their total weight. So, carbon atoms play a prominent role in living systems. They play a vital role in the construction of proteins. Unquestionably, carbon is the single most important element in our lives.

Of what are the proteins composed? Besides having lots of carbon, is there a common denominator among the myriads of different proteins found throughout the wide spectrum of cells alive today? The answer is yes, for experiments have shown that proteins are made of a rather small group of molecules, called amino acids. Although chemists have synthesized many such acids artificially, only 20 (plus two rare ones) of these structural units comprise the millions of proteins found in Earth's life—not just human life; all life. Amino acids are one of the two basic building blocks of life. Nucleotide Bases Mindful of the molecular structure of proteins, we return to our original concern: How are proteins made within organisms in order to keep them alive?

Specifically, what chemical process serves to combine amino acids in order to replenish the dead cytoplasm in all living systems? Whatever that process, it must be of central importance since the production of protein is absolutely vital to an organism's well-being—not any random collection of proteins, but exactly the right kinds of proteins, with their amino acids strung along in precisely the right order. To appreciate how protein is constructed with the correct numbering and sequencing, we defer to the nucleic acids, another of life's basic ingredients.

Nucleic acids, like proteins, are long chainlike groups of molecules, most of them also rich in carbon. Their name derives from the fact that these acids were first found in the biological nuclei of cells. Though chemists know of a large variety of them, the nucleic acids, again like proteins, are made of only a small number of key compounds. Called nucleotide bases, these are the second group of life's building blocks. The biochemical role of the bases is best illustrated by the most famous of all nucleic acids—deoxyribonucleic acid, alias DNA.

Most of the DNA molecule is made of a long string of 4 fundamental bases—adenine, cytosine, guanine, and thymine—that repeat over and over. A fifth nucleotide base, uracil, is used in the construction of other nucleic acids, though not in DNA. These 5 kinds of bases play much the same role for nucleic acids as do the 20 kinds of amino acids for proteins. Each nucleotide base is only slightly more complex than the amino acids, also being a molecular assemblage of carbon, hydrogen, oxygen, and nitrogen atoms—"CHON," for short. Parts of the bases bend around and attach to themselves in a ring, thereby becoming a little more stable.

Preservation of the exact structure of the original DNA molecule is the most important feature of replication. All the information about the specific duties of that type of cell—whether a blood cell, hair cell, muscle cell, or whatever—passes from an old cell to a newly created one. Accordingly, the biological function of the "daughter" cell remains identical to that of the "parent" cell. In this way, DNA molecules, whose functional units are the genes, are responsible for directing inheritance from generation to generation.

Just as for amino-acid sequences in proteins, the *order* of nucleotide bases as well as their *number* is paramount in the construction of nucleic acids. The sequence of bases along a nucleic-acid molecule specifies the physical and chemical behaviour of that particular gene. In turn, all the genes of a living system collectively comprise a genetic code—an encyclopedic compendium of the physical and chemical properties of all the system's cells and all its functions. In a very real sense, the two most important features of a living organism—its structure and its function—depend chiefly on the nucleic-acid molecules in the many nuclei of its cells, for these are the material entities that are passed on, or inherited, from one generation of cells to the next.

In analogy with another type of information storage—this Web site, for example—the individual bases can be considered words, the base pairs a sentence, and the whole DNA molecule a web of instructions. The words and sentences must be in the right order to give meaning to the Web site. An entire network of such instructional sites then comprises the genetic code for all the varied functions performed by any living organism. In short, a full set of DNA molecules is really information—a blueprint, or master plan, for every life form.

The nature of all living creatures is ultimately prescribed by the structure of their DNA molecules. These molecules specify, not only how one type of organism differs from another in both makeup and personality, but also how the physical and chemical events inside a cell properly coordinate so that the overall activity of the cell is as it ought to be. At first glance, it would seem impossible that one molecule could do all this—namely, dictate the behaviour of all the myriad life forms in the world today. After all, DNA has only 4 types of nucleotide bases. But DNA is the largest molecule known.

In advanced organisms such as vertebrates, a DNA molecule can have as many as 10^8 bases or 10^{10} separate atoms, making the molecule nearly a meter (three feet) long if extended end to end; in humans, ~2m of DNA are squeezed into every human cell, and if all the DNA in a single person were unwound it would stretch ~10^9 km, or several roundtrips between Sun and Earth.

In the above analogy where a DNA base equals a word, a single DNA molecule would resemble a 100-page manuscript. Consequently, huge numbers of possible combinations of bases guarantee a vast array of diverse living creatures, each with a different appearance, style, and personality. Yet, at the microscopic level, all creatures—without exception—are basically made of the same two dozen or so acids and bases, the very building blocks of life as we know it. The common molecular content pervading all life on Earth is our best evidence that every living thing dates back to a single-celled ancestor—the so-called LCA, or Last Common Ancestor—billions of years ago. Protein Synthesis Regarding our earlier query of protein synthesis, continuous production of the cytoplasm's proteins makes heavy use of the cell's nucleic acids. The sequence of events, typically goes as follows: Just prior to cell division, the DNA molecule sends a related RNA molecule out of the biological nucleus and into the cytoplasm.

RNA stands for ribonucleic acid—a smaller, single-stranded version of the normally double-stranded DNA (wherein, for RNA, the thymine nucleotide base is replaced by the uracil base). The RNA molecule acts like a messenger, carrying instructions from the DNA molecule. Once in the cytoplasm, single-stranded RNA attracts freely floating amino acids to its uncoupled bases. Only certain amino acids can successfully attach to the RNA bases, since the electromagnetic forces of RNA's bases attract some amino acids, while repelling others.

After some time—usually a few microseconds—the single-stranded RNA molecule fills its strand, partly by accidentally colliding and sticking, with an entire complement of proper amino acids. It's not all chance, however; these acids also attach to one another by means of their own electromagnetic forces, which are governed by physical law. Finally, when the long chain of amino acids is fully assembled along the entire length of the messenger RNA molecule, the chain detaches and drifts off into the cytoplasm of the cell.

A protein has thus formed. But, again, this is no ordinary protein created entirely at random. Rather, it's a specific protein formed according to the instructions provided by the RNA molecule, taking into account both chance (the random collisions) and necessity (the linking forces). In this way, RNA acts as a prescription or template on which protein molecules are built—a template originating in the cell nucleus with the DNA molecule itself.

This, then, is a highly simplified account of the way that proteins are continually replenished in living organisms. All living systems grow and eventually become biologically stabilized in this same way. The whole process is occurring repeatedly in our bodies right now. Of course, different organisms have different genes and therefore manufacture different proteins—except for identical twins, which do have the same DNA structure. In reality, life is much more complicated, as a single gene doesn't encode only one protein. One gene often yields a variety of proteins, which partly explains why humanity has hundreds of thousands of different proteins despite the so-called human genome—

the sum of all our genes—containing only ~30,000 genes. Genes and proteins: The first directs reproduction, the passage of heredity from one generation of life to the next. The second directs metabolism, the daily flow of incoming food (which is high-grade energy) and outgoing wastes (low-grade energy). While genes surely contain the recipes for making proteins, it's the proteins that comprise the (structural) bricks and motor of cells and that do most of the (functional) work. Whether in man, mouse, or microbe, the genes mastermind life and the proteins maintain its well being.

Life's Handedness

A central puzzle in modern biochemistry is life's chirality—that is, a tendency for life's molecules to have a certain preferential orientation, or "handedness." Much of life is said to be inherently left-handed, especially its amino acids. No one has ever been able to explain satisfactorily how life became so asymmetric. Yet broken symmetry seems as central to biology and life on Earth as it is to physics and the behaviour of matter in the early Universe. Asymmetry may well be an essential prerequisite for the origin and evolution of complexity throughout all of Nature.

A molecule's orientation can be determined by watching the behaviour of polarized light passing through it. This is a type of light having an aligned plane of vibration, which can rotate right or left when encountering a material substance. Key molecules of life—especially the usual 20 amino acids comprising the structure of all proteins—are almost exclusively of the left-handed variety, since light moving through them rotates left. By contrast, the nucleotide bases and sugars that comprise RNA and DNA tend to be right-handed. "Handedness" is one of the most striking properties of life on Earth. But we don't understand it.

Life's amino-acid preference for left-handedness is particularly puzzling because such molecules, when artificially produced in the laboratory, invariably show an equal mixture of left- and right-handed configurations. Furthermore, should a right-handed amino acid drift into a living organism, the catalysts that control protein production will quickly destroy it. Not only that, when a living organism dies and decays, thermal fluctuations change molecular shapes randomly, so that eventually an even left-right mixture results. Why terrestrial life employs only left-handed amino acids (or right-handed nucleic acids) is one of the great unsolved mysteries of chemical evolution. One possibility is that the first organism just happened, simply by chance, to be left-handed. If life arose only once on Earth, all its descendants would then also be left-handed; the continuity of life is a mere copying process. An alternative, less chancy possibility is that both left- and right-handed organisms originated, perhaps on different occasions billions of years ago, but that left-handed life proved advantageous in eliminating all competitors.

The ability to make an extra amino acid or a healthy vitamin, for instance, might have provided such an advantage. If minerals acted as catalytic templates for life's origin, as appraised later in this CHEMICAL EPOCH, some crystals might have attracted left- and right-handed amino acids differently. For example, the rock calcite (a common mineral, $CaCO_3$, that forms limestone and marble) does display this kind of asymmetric property, possibly acting as a determined selector and not a purely chance event, that could explain why much of life is preferentially left-handed.

Yet another intriguing idea interfaces physics with biology. In brief, life's handedness might result from one of Nature's basic forces, once again invoking determinism more than chance. The weak nuclear force operates on size scales smaller than nuclear dimensions and is thus often dismissed as unimportant to atomic physics, let alone molecular biology. However, as noted in the earlier PARTICLE EPOCH, the weak force has now been merged with the electromagnetic force, which biologists often refer to as the "life force."

And since some weak-interaction events studied in nuclear laboratories do show a preference for one handedness over the other (more elementary particles spin clockwise than counterclockwise during weak-force radioactive events), there might well be a very small (thus far undetected) difference in total energy between left- and right-handed molecules. If true, the left-handed amino acids could have been advantageously selected, as this is the lower-energy state preferred by Nature.

Chapter 10

Chemical Environment of Cell

The living cell is a symphony of thousands of chemical reactions all miraculously timed and coordinated to perform all the functions necessary for life. Amazingly, this symphony has only a few major players; only six elements carbon, hydrogen, oxygen, nitrogen, phosphorus, and sulfur (sometimes called CHNOPS;>) make up about 98% of the mass of all living organisms. Carbon is a unique element with the remarkable ability to form strong, stable chemical bonds with other atoms (keeps you from falling apart.) Each carbon atom can form four bonds with other atoms. (Sometimes, two atoms will form more than one bond between themselves making a double bond or even a triple bond.)

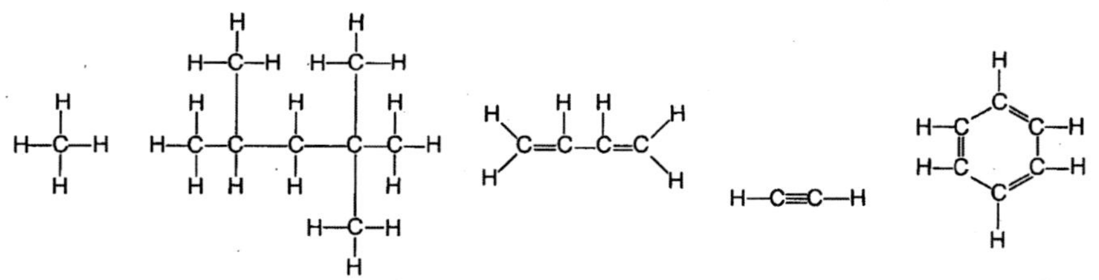

Methane **Iso-Octane** **Butadiene** **Acetylene** **Benzene**

This bonding ability allows carbon atoms to form chains of almost unlimited length. These chains can be closed on themselves to form rings or may branch wildly. This gives great variety to the kinds of molecules that carbon can form. Below are just a few examples of the many ways carbon chains can be arranged to form the skeleton for different molecules. Atoms of hydrogen and oxygen and less frequently nitrogen, sulfur, or phosphorous are bonded to the carbon skeleton to form giant molecules called macromolecules. A few other elements may occur in trace amounts.

The four major types of macromolecules found in living cells—carbohydrates, lipids, proteins, and nucleic acids—are made of smaller, repeating subunits called monomers. The monomers are not always identical but they always have similar chemical structures. They are joined together by a series of chemical reactions in a process called polymerization to form large, complex molecules called polymers.

The Four Major Types of Macromolecules Found in Living Cells				
Macromolecule	Elements	Monomer	Polymer	example
Carbohydrate	C, H, O	Simple sugars	Polysaccharide	Starch
Lipids	C, H, O	Fatty acids & glycerol	Lipid	Fats, oils, waxes
Proteins	C, H, O, N, S	Amino acids	Polypeptides	Insulin
Nucleic acids	C, H, O, P	Nucleotides	Nucleic acids	DNA

The chemical diversity that polymerization allows living things is similar to the diversity that our alphabet allows our language. Although there are only 26 letters in our English alphabet, our ability to join them together to form words gives us an almost infinite variety of possible words.

Similarly, the monomer units of macromolecules can be arranged with an almost endless potential for variety. The functions of macromolecules are directly related to their shapes and to the chemical properties of their monomers. The way the monomers are arranged in the macromolecule determines its shape and its function in the same way that the arrangement of the letters in a word determine its sound and meaning.

Much of a cell's activities involve the arranging and rearranging and bonding of macromolecules. It is the job of DNA both directly and indirectly to coordinate and direct these activities. An understanding of the structure and functions of carbohydrates and lipids is not particularly key to the understanding of molecular genealogy. However, it may be helpful to take a quick look at the structure and function of proteins before moving on to the nitty gritty of the nucleic acids (of which DNA is one.)

Nucleic Acids

Nucleic acids are nucleotide polymers (from the Greek word poly, meaning "several", and mer, meaning "unit"), that store and transmit genetic information. Only 4 different nucleotides are used in nucleic acid biosynthesis. Genetic information contained in nucleic acids is stored and replicated in chromosomes, which contain genes (from the Greek word gennan, meaning "to produce"). A chromosome is a deoxyribonucleic acid (DNA) molecule, and genes are segments of intact DNA. The total number of genes in any given mammalian cell may total several thousand.

When a cell replicates itself, identical copies of DNA molecules are produced, therefore the hereditary line of descent is conserved, and the genetic information carried on DNA is available to direct the occurrence of virtually all chemical reactions within the cell. The bulk of genetic information carried on DNA provides instructions for the assembly of virtually every protein molecule within the cell. The flow of information from nucleic acids to protein is commonly represented as DNA —> messenger ribonucleic acid (mRNA) —> transfer RNA (tRNA) —> ribosomal RNA (rRNA) —> protein, which indicates that the nucleotide sequence in a gene of DNA specifies the assembly of a nucleotide sequence in an mRNA molecule, which in turn directs the assembly of the amino acid sequence in protein through a tRNA and rRNA molecules.

Proteins

Proteins are amino acid polymers responsible for implementing instructions contained within the genetic code. Twenty different amino acids are used to synthesize proteins, about half are formed as metabolic intermediates, while the remainder must be provided through the diet. The latter group is referred to as "essential" amino acids.

Each protein formed in the body, unique in its own structure and function, participates in processes that characterize the individuality of cells, tissues, organs, and organ systems. A typical cell contains thousands of different proteins, each with a different function, and many serve as enzymes that catalyse (or speed) reactions. Virtually every reaction in a living cell requires an enzyme. Other proteins transport different compounds either outside or inside cells {e.g., lipoproteins and transferring (an iron-binding protein) in plasma, or bilirubinbinding proteins in liver cells}; some act as storage proteins (e.g., myoglobin binds and stores O2 in muscle cells); others as defence proteins in blood or on the surface of cells (e.g., clotting proteins and immunoglobulins); others as contractile proteins (e.g., the actin, myosin and troponin of skeletal muscle fibers); and others are merely structural in nature (e.g., collagen and elastin). Proteins, unlike glycogen and triglyceride, are usually not synthesized and stored as nonfunctional entities.

Polysaccharides

Polysaccharides are polymers of simple sugars (i.e., monosaccharides). (The term saccharide is derived from the Greek word sakchar, meaning "sugar or sweetness".) Some polysaccharides are homogeneous polymers that contain only one kind of sugar (e.g., glycogen), while others are complex heterogenous polymers that contain 8-10 types of sugars.

In contrast to heterogenous polymers (e.g., proteins, nucleic acids, and some polysaccharides), homogenous polymers are considered to be "noninformational". Polysaccharides, therefore, can occur as functional and structural components of cells (e.g., glycoproteins and glycolipids), or merely as noninformational storage forms of energy (e.g., glycogen). The 8-10 monosaccharides that become the building blocks for heterogenous polysaccharides can be synthesized from glucose, or formed from other metabolic intermediates.

Lipids

Lipids (from the Greek word lipos, meaning "fat") are naturally occurring, nonpolar substances that are mostly insoluble in water (with the exceptions being the short-chain volatile fatty acids and ketone bodies), yet soluble in nonpolar solvents (like chloroform and ether). They serve as membrane components (cholesterol, glycolipids and phospholipids), storage forms of energy (triglycerides), precursors to other important biomolecules (fatty acids), insulation barriers (neutral fat stores), protective coatings to prevent infection and excessive gain or loss of water, and some vitamins (A, D, E, and K) and hormones (steroid hormones).

Major classes of lipids are the saturated and unsaturated fatty acids (short, medium, and long-chain), triglycerides, lipoproteins {i.e., chylomicrons (CMs), very low density (VLDL), low density (LDL), intermediate density (IDL), and high density lipoproteins (HDL)}, phospholipids and glycolipids, steroids (cholesterol, progesterone, etc.), and eicosanoids (prostaglandins, thromboxanes, and leukotrienes). All lipids can be synthesized from acetyl-CoA, which in turn can be generated from numerous different sources, including carbohydrates, amino acids, short-chain volatile fatty acids (e.g., acetate), ketone bodies, and fatty acids. Simple lipids include only those that are esters of fatty acids and an alcohol (e.g., mono-, di- and triglycerides).

Compound lipids include various materials that contain other substances in addition to an alcohol and fatty acid (e.g., phosphoacylglycerols, sphingomyelins, and cerebrosides), and derived lipids include those that cannot be neatly classified into either of the above (e.g., steroids, eicosanoids, and the fat-soluble vitamins).

Although the study of physiological chemistry emphasizes organic molecules, the inorganic elements (sometimes subdivided into macrominerals, trace elements, and ultra trace elements), are also important. Several are "essential" nutrients, and therefore like certain amino acids and unsaturated fatty acids, must be supplied in the diet.

Inorganic elements are typically present in cells as ionic forms, existing as either free ions or complexed with organic molecules. Many "trace elements" are known to be essential for life, health, and reproduction, and have well-established actions (e.g., cofactors for enzymes, sites for binding of oxygen (in transport), and structural components of nonenzymatic macromolecules. Some investigators have speculated that perhaps all of the elements on the periodic chart will someday by shown to exhibit physiologic roles in mammalian life.

Because life depends upon chemical reactions, and because most all diseases in animals are manifestations of abnormalities in biomolecules, chemical reactions, or biochemical pathways, physiological chemistry has become the language of all basic medical sciences. A fundamental understanding of this science is therefore needed not only to help illuminate the origin of disease, but also to help formulate appropriate therapies.

The Cell Membrane

A thin membrane surrounds every living cell, delimiting the cell from the environment around it. Enclosed by this cell membrane (also known as the plasma membrane) are the cell's constituents, often large, water-soluble, highly charged molecules such as proteins, nucleic acids, carbohydrates, and substances involved in cellular metabolism. Outside the cell, in the surrounding water-based environment, are ions, acids, and alkalis that are toxic to the cell, as well as nutrients that the cell must absorb in order to live and grow. The cell membrane, therefore, has two functions: first, to be a barrier keeping the constituents of the cell in and unwanted substances out and, second, to be a gate allowing transport into the cell of essential nutrients and movement from the cell of waste products.

Chemical Composition and Membrane Structure

Most current knowledge about the biochemical constituents of cell membranes originates in studies of red blood cells. The chief advantage of these cells for experimental purposes is that they may be obtained easily in large amounts and that they have no internal membranous organelles to interfere with study of their cell membranes. Careful studies of these and other cell types have shown that all membranes are composed of proteins and fatty-acid-based lipids. Membranes actively involved in metabolism contain a higher proportion of protein; thus, the membrane of the mitochondrion, the most rapidly metabolizing organelle of the cell, contains as much as 75 percent protein, while the membrane of the Schwann cell, which forms an insulating sheath around many nerve cells, has as little as 20 percent protein.

Membrane Lipids

Membrane lipids are principally of two types, phospholipids and sterols (generally cholesterol). Both types share the defining characteristic of lipids—they dissolve readily in organic solvents—but in addition they both have a region that is attracted to and soluble in water. This "amphiphilic" property (having a dual attraction; i.e., containing both a lipid-soluble and a water-soluble region) is basic to the role of lipids as building blocks of cellular membranes.

Phospholipid molecules have a head (often of glycerol) to which are attached two long fatty acid chains that look much like tails. These tails are repelled by water and dissolve readily in organic solvents, giving the molecule its lipid character. To another part of the head is attached a phosphoryl group with a negative electrical charge; to this group in turn is attached another group with a positive or neutral charge. This portion of the phospholipid dissolves in water, thereby completing the molecule's amphiphilic character. In contrast, sterols have a complex hydrocarbon ring structure as the lipid-soluble region and a hydroxyl grouping as the water-soluble region.

When dry phospholipids, or a mixture of such phospholipids and cholesterol, are immersed in water under laboratory conditions, they spontaneously form globular structures called liposomes. Investigation of the liposomes shows them to be made of concentric spheres, one sphere inside of another and each forming half of a bilayered wall.

A bilayer is composed of two sheets of phospholipid molecules with all of the molecules of each sheet aligned in the same direction. In a water medium, the phospholipids of the two sheets align so that their water-repellent, lipid-soluble tails are turned and loosely bonded to the tails of the molecules on the other sheet. The water-soluble heads turn outward into the water, to which they are chemically attracted. In this way, the two sheets form a fluid, sandwichlike structure, with the fatty acid chains in the middle mingling in an organic medium while sealing out the water medium.

This type of lipid bilayer, formed by the self-assembly of lipid molecules, is the basic structure of the cell membrane. It is the most stable thermodynamic structure that a phospholipid-water mixture can take up: the fatty acid portion of each molecule dissolved in the organic phase formed by the identical regions of the other molecules and the water-attractive regions surrounded by water and facing away from the fatty acid regions. The chemical affinity of each region of the amphiphilic molecule is thus satisfied in the bilayer structure.

Membrane Proteins

Membrane proteins are also of two general types. One type, called the extrinsic proteins, is loosely attached by ionic bonds or calcium bridges to the electrically charged phosphoryl surface of the bilayer. They can also attach to the second type of protein, called the intrinsic proteins. The intrinsic proteins, as their name implies, are firmly embedded within the phospholipid bilayer. Almost all intrinsic proteins contain special amino acid sequences, generally about 20- to 24-amino acids long, that extend through the internal regions of the cell membrane. Most intrinsic and extrinsic proteins bear on their outer surfaces side chains of complex sugars, which extend into the aqueous environment around the cell. For this reason, these proteins are often referred to as glycoproteins. Some glycoproteins are involved in cell-to-cell recognition.

Membrane Fluidity

One of the triumphs of cell biology during the decade from 1965 to 1975 was the recognition of the cell membrane as a fluid collection of amphiphilic molecules. This array of proteins, sterols, and phospholipids is organized into a liquid crystal, a structure that lends itself readily to rapid cell growth. Measurements of the membrane's viscosity show it as a fluid one hundred times as viscous as water, similar to a thin oil. The phospholipid molecules diffuse readily in the plane of the bilayer. Many of the membrane's proteins also have this freedom of movement, but some are fixed in the membrane by interaction with the cell's cytoskeleton. Newly synthesized phospholipids insert themselves easily into the existing cell membrane. Intrinsic proteins are inserted during their synthesis on ribosomes bound to the endoplasmic reticulum, whereas extrinsic proteins found on the internal surface of the cell membrane are synthesized on free, or unattached, ribosomes, liberated into the cytoplasm, and then brought to the membrane.

Transport Across the Membrane

The chemical structure of the cell membrane makes it remarkably flexible, the ideal boundary for rapidly growing and dividing cells. Yet the membrane is also a formidable barrier, allowing some dissolved substances, or solutes, to pass while blocking others. Lipid-soluble molecules and some small molecules can permeate the membrane, but the lipid bilayer effectively repels the many large, water-soluble molecules and electrically charged ions that the cell must import or export in order to live. Transport of these vital substances is carried out by certain classes of intrinsic proteins that form a variety of transport systems: some are open channels, which allow ions to diffuse directly into the cell; others are "facilitators," which, through a little-understood chemical transformation, help solutes diffuse past the lipid screen; yet others are "pumps," which force solutes through the membrane when they are not concentrated enough to diffuse spontaneously. Particles too large to be diffused or pumped are often swallowed or disgorged whole by an opening and closing of the membrane.

Behind this movement of solutes across the cell membrane is the principle of diffusion. According to this principle, a dissolved substance diffuses down a concentration gradient; that is, given no energy from an outside source, it moves from a place where its concentration is high to a place where its concentration is low. Diffusion continues down this gradually decreasing gradient until a state of equilibrium is reached, at which point there is an equal concentration in both places and an equal, random diffusion in both directions. A solute at high concentration is at high free energy; that is, it is capable of doing more "work" (the work being that of diffusion) than a solute at low concentration. In performing the work of diffusion, the solute loses free energy, so that, when it reaches equilibrium at a lower concentration, it is unable to return spontaneously (under its own energy) to its former high concentration. However, by the addition of energy from an outside source (through the work of an ion pump, for example), the solute may be returned to its former concentration and state of high free energy. This "coupling" of work processes is, in effect, a transferal of free energy from the pump to the solute, which is then able to repeat the work of diffusion.

For most substances of biological interest, the concentrations inside and outside the cell are different, creating concentration gradients down which the solutes spontaneously diffuse, provided they can permeate the lipid bilayer. Membrane channels and diffusion facilitators bring them through the membrane by passive transport; that is, the changes that the proteins undergo in order to facilitate diffusion are powered by the diffusing solutes themselves. For the healthy functioning of the cell, certain solutes must remain at different concentrations on each side of the membrane; if through diffusion they approach equilibrium, they must be pumped back up their gradients by the process of active transport. Those membrane proteins serving as pumps accomplish this by coupling the energy required for transport to the energy produced by cell metabolism or by the diffusion of other solutes.

Permeation

Permeation is the diffusion, through a barrier, of a substance in solution. The rates at which biologically important molecules cross the cell membrane through permeation vary over an enormous range. Proteins and sugar polymers do not permeate at all; in contrast, water and alcohols permeate most membranes in less than a second. This variation, caused by the lipid bilayer, gives the membrane its characteristic permeability. Permeability is measured as the rate at which a particular substance in solution crosses the membrane.

For all cell membranes that have been studied in the laboratory, permeability increases in parallel with the permeant's ability to dissolve in organic solvents. The consistency of this parallel has led researchers to conclude that permeability is a function of the fatty acid interior of the lipid bilayer, rather than its phosphoryl exterior. This property of dissolving in organic solvents rather than water is given a unit of measure called the partition coefficient. The greater the solubility of a substance, the higher its partition coefficient, and the higher the partition coefficient, the higher the permeability of the membrane to that particular substance. For example, the water solubility of hydroxyl, carboxyl, and amino groups reduces their solubility in organic solvents and, hence, their partition coefficients. Cell membranes have been observed to have low permeability toward these groups. In contrast, lipid-soluble methyl residues and hydrocarbon rings, which have high partition coefficients, penetrate cell membranes more easily—a property useful in designing chemotherapeutic and pharmacological drugs.

For two molecules of the same partition coefficient, the one of greater molecular weight, or size, will in general cross the membrane more slowly. In fact, even molecules with very low partition coefficients can penetrate the membrane if they are small enough. Water, for example, is insoluble in organic solvents, yet it permeates cell membranes because of the small size of its molecules. The size selectivity of the lipid bilayer is a result of its being not a simple fluid, the molecules of which move around and past a diffusing molecule, but an organized matrix, a kind of fixed grate, composed of the fatty acid chains of the phospholipids through which the diffusing molecule must fit. Many substances do not actually cross the cell membrane through permeation of the lipid bilayer. Some electrically charged ions, for example, are repelled by organic solvents and therefore cross cell membranes with great difficulty, if at all. In these cases special holes in the membrane, called channels, allow specific ions and small molecules to diffuse directly through the bilayer.

Membrane Channels

Biophysicists measuring the electric current passing through cell membranes have found that, in general, cell membranes have a vastly greater electrical conductance than does a membrane bilayer composed only of phospholipids and sterols. This greater conductance is thought to be conferred by the cell membrane's proteins. A current flowing across a membrane often appears on a recording instrument as a series of bursts of various heights. These bursts represent current flowing through open channels, which are merely holes formed by intrinsic proteins traversing the lipid bilayer. No significant current flows through the membrane when no channel is open; multiple bursts are recorded when more than one channel is open.

A rich variety of channels has been isolated and analysed from a wide range of cell membranes. Invariably intrinsic proteins, they contain numerous amino acid sequences that traverse the membrane, clearly forming a specific hole, or pore. Certain channels open and close spontaneously. Some are gated, or opened, by the chemical action of a signalling substance such as calcium, acetylcholine, or glycine, whereas others are gated by changes in the electrical potential across the membrane. Channels may possess a narrow specificity, allowing passage of only potassium or sodium, or a broad specificity, allowing passage of all positively charged ions (cations) or of all negatively charged ions (anions). There are channels called gap junctions that allow the passage of molecules between pairs of cells.

The gating of channels with a capacity for ion transport is the basis of the many nerve-nerve, nerve-muscle, and nerve-gland interactions underlying neurobiological behaviour. These actions depend on the electric potential of the cell membrane, which varies with the prevailing constituents in the cell's environment. For example, if a channel that admits only potassium ions is present in a membrane separating two different potassium chloride solutions, the positively charged potassium ions tend to flow down their concentration gradient through the channel. The negatively charged chloride ions remain behind. This separation of electric charges sets up an electric potential across the membrane called the diffusion potential. The size of this potential depends on, among other factors, the difference in concentrations of the permeating ion across the membrane. The cell membrane in general contains the channels of widely different ion specificities, each channel contributing to the overall membrane potential according to the permeability and concentration ratio of the ion passing through it. Since the channels are often gated, the membrane's potential is determined by which channels are open; this in turn depends on the concentrations of signalling molecules and may change with time according to the membrane potential itself.

Most cells have about a tenfold higher concentration of sodium ions outside than inside and a reverse concentration ratio of potassium ions. Free calcium ions can be 10,000 times more concentrated outside the cell than inside. Thus, sodium-, potassium-, and calcium-selective membrane channels, by allowing the diffusion of those ions past the cell membrane and causing fluctuations in the membrane's electric potential, frequently serve as transmitters of signals from nerve cells. Ion diffusion threatens to alter the concentration of ions necessary for the cell to function. The proper distribution of ions is restored by the action of ion pumps.

Facilitated Diffusion

Many water-soluble molecules that cannot penetrate the lipid bilayer are too large to fit through open channels. In this category are sugars and amino acids. Some ions too do not diffuse through channels. These vital substances enter and leave the cell through the action of membrane transporters, which, like channels, are intrinsic proteins that traverse the cell membrane. Unlike channels, transporter molecules do not simply open holes in the membrane. Rather, they present sites on one side of the membrane to which molecules bind through chemical attraction. The binding site is highly specific, often fitting the atomic structure of only one type of molecule. When the molecule has attached to the binding site, then, in a process not fully understood, the transporter brings it through the membrane and releases it on the other side.

This action is considered a type of diffusion because the transported molecules move down their concentration gradients, from high concentration to low. To activate the action of the transporter, no other energy is needed than that of the chemical binding of the transported molecules. This action upon the transporter is similar to catalysis, except that the molecules (in this context called substrates) catalyse not a chemical reaction but their own translocation across the cell membrane. Two such substrates are glucose and the bicarbonate ion.

The Glucose Transporter

This sugar-specific transport system enables half of the glucose present inside the cell to leave within four seconds at normal body temperature. The glucose transporter is clearly not a simple membrane channel. First, unlike a channel, it does not select its permeants by size, as one type of glucose is observed to move through the system a thousand times faster than its identically sized optical isomer. Second, it operates much more slowly than do most channels, moving only 1,000 molecules per second while a channel moves 1,000,000 ions. The most important difference between a membrane channel and the glucose transporter is the conformational change that the transporter undergoes while moving glucose across the membrane. Alternating between two conformations, it moves its glucose-binding site from one side of the membrane to the other. By "flipping" between its two conformational states, the transporter facilitates the diffusion of glucose; that is, it enables glucose to avoid the barrier of the cell membrane while moving spontaneously down its concentration gradient. When the concentration reaches equilibrium, net movement of glucose ceases. A facilitated diffusion system for glucose is present in many cell types. Similar systems transporting a wide range of other substrates (e.g., different sugars, amino acids, nucleosides, and ions) are also present.

The Anion Transporter

The best-studied of the facilitated diffusion systems is that which catalyzes the exchange of anions across the red blood cell membrane. The exchange of hydroxyl for bicarbonate ions, each ion simultaneously being moved down its concentration gradient in opposite directions by the same transport molecule, is of great importance in enhancing the blood's capacity to carry carbon dioxide from tissues to the lungs. The exchange molecule for these anions is the major intrinsic protein of red blood cells; one million of them are present on each cell, the polypeptide chain of each molecule traversing the membrane at least six times.

Secondary Active Transport

In some cases the problem of forcing a substrate up its concentration gradient is solved by coupling that upward movement to the downward flow of another substrate. In this way the energy-expending diffusion of the driving substrate powers the energy-absorbing movement of the driven substrate from low concentration to high. Because this type of active transport is not powered directly by the energy released in cell metabolism, it is called secondary. There are two kinds of secondary active transport: counter-transport, in which the two substrates cross the membrane in opposite directions, and cotransport, in which they cross in the same direction.

Counter-transport

An example of this system (also called antiport) begins with the sugar transporter. There are equal concentrations of glucose on both sides of the cell. A high concentration of galactose is then added outside the cell. Galactose competes with glucose for binding sites on the transport protein, so that mostly galactose—and a little glucose—enter the cell. The transporter itself, undergoing a conformational change, presents its binding sites for sugar at the inner face of the membrane. Here, at least transiently, glucose is in excess of galactose; it binds to the transporter and leaves the cell as the transporter switches back to its original conformation. Thus, glucose is pumped out of the cell against its gradient in exchange for the galactose riding into the cell down its own gradient.

Many counter-transport systems operate across the cell membranes of the body. A well-studied system (present in red blood cells, nerve cells, and muscle cells) pumps one calcium ion out of the cell in exchange for two or three sodium ions. This system helps maintain the low calcium concentration required for effective cellular activity. A different system, present in kidney cells, counter-transports hydrogen ions and sodium ions in a one-for-one ratio. This is important in stabilizing acidity by transporting hydrogen ions out of the body as needed.

Co-transport

In co-transport (sometimes called symport) two species of substrate, generally an ion and another molecule or ion, must bind simultaneously to the transporter before its conformational change can take place. As the driving substrate is transported down its concentration gradient, it drags with it the driven substrate, which is forced to move up its concentration gradient. The transporter must be able to undergo a conformational change when not bound to either substrate, so as to complete the cycle and return the binding sites to the side from which driving and driven substrates both move.

Sodium ions are usually the driving substrates in the co-transport systems of animal cells, which maintain high concentrations of these ions through primary active transport. The driven substrates include a variety of sugars, amino acids, and other ions. During the absorption of nutrients, for example, sugars and amino acids are removed from the intestine by co-transport with sodium ions. After passing across the glomerular filter in the kidney, these substrates are returned to the body by the same system. Plant and bacterial cells usually use hydrogen ions as the driving substrate; sugars and amino acids are the most common driven substrates. When the bacterium *Escherichia coli* must metabolize lactose, it co-transports hydrogen ions with lactose (which can reach a concentration 1,000 times higher than that outside the cell).

Primary Active Transport

The Sodium-potassium Pump

Human red blood cells contain a high concentration of potassium and a low concentration of sodium, yet the plasma bathing the cells is high in sodium and low in potassium. When whole blood is stored cold under laboratory conditions, the cells lose potassium and gain sodium until the concentrations across the membrane for both ions are at equilibrium. When the cells are restored to body temperature and given appropriate nutrition, they extrude sodium and take up potassium, transporting both ions against their respective gradients until the previous high concentrations are reached. This ion pumping is linked directly to the hydrolysis of adenosine triphosphate (ATP), the cell's repository of metabolic energy. For every molecule of ATP split, three ions of sodium are pumped out of the cell and two of potassium are pumped in.

An enzyme called sodium-potassium-activated ATPase has been shown to be the sodium-potassium pump, the protein that transports the ions across the cell membrane while splitting ATP. Widely distributed in the animal kingdom and always associated with the cell membrane, this ATPase is found at high concentration in cells that pump large amounts of sodium (e.g., in mammalian kidneys, in salt-secreting glands of marine birds, and in the electric organs of eels).

The enzyme, an intrinsic protein, exists in two major conformations whose interconversion is driven by the splitting of ATP or by changes in the transmembrane flows of sodium and potassium. When only sodium is present in the cell, the inorganic phosphate split from ATP during hydrolysis is transferred to the enzyme.

Release of the chemically bound phosphate from the enzyme is catalysed by potassium. Thus, the complete action of ATP splitting has been demonstrated to require both sodium (to catalyse the transfer of the phosphate to the enzyme) and potassium (to catalyse the release of the phosphate and free the enzyme for a further cycle of ATP splitting). Apparently, only after sodium has catalysed the transferal of the phosphate to the enzyme can it be transported from the cell. Similarly, only after potassium has released the phosphate from the enzyme can it be transported into the cell. This overall reaction, completing the cycle of conformational changes in the enzyme, involves a strict coupling of the splitting of ATP with the pumping of sodium and potassium. It is this coupling that creates primary active transport.

The sodium-potassium pump extrudes one net positive charge during each cycle of ATP splitting. This flow of current induces an electric potential across the membrane that adds to the potentials brought about by the diffusion of ions through gated channels. The pump's contribution to the overall potential is important in certain specialized nerve cells.

Calcium Pumps

Many animal cells can perform a primary active transport of calcium out of the cell, developing a 10,000-fold gradient of that ion. Calcium-activated ATPases have been isolated and shown to be intrinsic proteins straddling the membrane and undergoing conformational changes similar to those of the sodium-potassium-activated ATPase.

When a rise in the concentration of cellular calcium results from the opening of calcium-selective channels, the membrane's calcium pumps restore the low concentration.

Hydrogen Ion Pumps

Hydrochloric acid is produced in the stomach by the active transport of hydrogen ions from the blood across the stomach lining, or gastric mucosa. Hydrogen concentration gradients of nearly one million can be achieved by a hydrogen-potassium-activated ATP-splitting intrinsic protein in the cells lining the stomach. Apart from its specific ion requirements, the properties of this enzyme are remarkably similar to those of the sodium-potassium-activated enzyme and the calcium-activated enzyme. Other hydrogen-pumping ATP-splitting primary active transporters occur in intracellular organelles, in bacteria, and in plant cells. The steep gradient of hydrogen ions represents a store of energy that can be harnessed to the accumulation of nutrients or, in the case of bacterial flagella, to the powering of cell movement.

Transport of Particles

In bringing about transmembrane movements of large molecules, the cell membrane itself undergoes concerted movements during which part of the fluid medium outside of the cell is internalized (endocytosis) or part of the cell's internal medium is externalized (exocytosis). These movements involve a fusion between membrane surfaces, followed by the re-formation of intact membranes.

Endocytosis

In this process the cell membrane engulfs portions of the external medium, forms an almost complete sphere around it, and then draws the membrane-bounded vesicle, called an endosome, into the cell. Several types of endocytosis have been distinguished: in pinocytosis, the vesicles are small and contain fluid; in phagocytosis, the vesicles are larger and contain solid matter; and in receptor-mediated endocytosis, material binds to a specific receptor on the external face of the cell membrane, triggering the process by which it is engulfed. Cholesterol enters cells by the last route.

Exocytosis

In exocytosis, material synthesized within the cell that has been packaged into membrane-bound vesicles is exported from the cell following the fusion of the vesicles with the external cell membrane.

The materials so exported are cell-specific protein products, neurotransmitters, and a variety of other molecules.

Internal Membranes

The presence of internal membranes distinguishes eukaryotic cells (cells with a nucleus) from prokaryotic cells (those without a nucleus). Prokaryotic cells are small (one to five micrometres in length) and contain only a single cell membrane; metabolic functions are often confined to different patches of the membrane rather than to areas in the body of the cell. Typical eukaryotic cells, by contrast, are much larger, the cell membrane constituting only 10 percent or less of the total cellular membrane. Metabolic functions in these cells are carried out in the organelles, compartments sequestered from the cell body, or cytoplasm, by internal membranes.

This section discusses internal membranes as structural and functional components in the organelles and vesicles of eukaryotic cells. The principal organelles—the nucleus, mitochondrion, and (in plants) chloroplast—are discussed elsewhere. Of the remaining organelles, the lysosomes, peroxisomes, and (in plants) glyoxysomes enclose extremely reactive by-products and enzymes. Internal membranes form the mazelike endoplasmic reticulum, where cell membrane proteins and lipids are synthesized, and they also form the stacks of flattened sacs called the Golgi apparatus, which is associated with the transport and modification of lipids, proteins, and carbohydrates. Finally, internal cell membranes can form storage and transport vesicles and the vacuoles of plant cells. Each membrane structure has its own distinct composition of proteins and lipids enabling it to carry out unique functions.

General Functions and Characteristics

Like the cell membrane, membranes of some organelles contain transport proteins, or permeases, that allow chemical communication between organelles. Permeases in the lysosomal membrane, for example, allow amino acids generated inside the lysosome to cross into the cytoplasm, where they can be used for the synthesis of new proteins. Communication between organelles is also achieved by the membrane budding processes of endocytosis and exocytosis, which are essentially the same as in the cell membrane. On the other hand, the biosynthetic and degradative processes taking place in different organelles may require conditions greatly different from those of other organelles or of the cytosol (the fluid part of the cell surrounding the organelles). Internal membranes maintain these different conditions by isolating them from one another. For example, the internal space of lysosomes is much more acidic than that of the cytosol—pH 5 as opposed to pH 7—and is maintained by specific proton-pumping transport proteins in the lysosome membrane.

Another function of organelles is to prevent competing enzymatic reactions from interfering with one another. For instance, essential proteins are synthesized on the rough endoplasmic reticulum and in the cytosol, while unwanted proteins are broken down in the lysosomes and also, to some extent, in the cytosol. Similarly, fatty acids are made in the cytosol and then either broken down in the mitochondria for the synthesis of ATP or degraded in the peroxisomes with concomitant generation of heat. These processes must be kept isolated. Organelle membranes also prevent potentially lethal by-products or enzymes from attacking sensitive molecules in other regions of the cell by sequestering such degradative activities in their respective membrane-bounded compartments.

The internal membranes of eukaryotic cells differ both structurally and chemically from the outer cell membrane. Like the outer membrane, they are constructed of a phospholipid bilayer into which are embedded, or bound, specific membrane proteins.

The three major lipids forming the outer membrane—phospholipids, cholesterol, and glycolipids—are also found in the internal membranes, but in different concentrations. Phospholipid is the primary lipid forming all cellular membranes. Cholesterol, which contributes to the fluidity and stability of all membranes, is found in internal membranes at about 25 percent of the concentration in the outer membrane. Glycolipids are found only as trace components of internal membranes, whereas they constitute approximately 5 percent of the outer membrane lipid.

Cellular Organelles and Their Membranes

The Vacuole

Most plant cells contain one or more membrane-bound vesicles called vacuoles. Within the vacuole is the cell sap, a water solution of salts and sugars kept at high concentration by the active transport of ions through permeases in the vacuole membrane. Proton pumps also maintain high concentrations of protons in the vacuole interior. These high concentrations cause the entry, via osmosis, of water into the vacuole, which in turn expands the vacuole and generates a hydrostatic pressure, called turgor, that presses the cell membrane against the cell wall. Turgor is the cause of rigidity in living plant tissue. In the mature plant cell, as much as 90 percent of cell volume may be taken up by a single vacuole; immature cells typically contain several smaller vacuoles.

Microbodies

Microbodies are roughly spherical in shape, bound by a single membrane, and are usually 0.5 to 1 micrometre in diameter. There are several types, by far the most common of which is the peroxisome. Peroxisomes derive their name from hydrogen peroxide, a reactive intermediate in the process of molecular breakdown that occurs in the microbody. Peroxisomes contain type II oxidases, which are enzymes that use molecular oxygen in reactions to oxidize organic molecules. A product of these reactions is hydrogen peroxide, which is further metabolized into water and oxygen by the enzyme catalase, a predominant constituent of peroxisomes. In addition, peroxisomes contain other enzyme systems that degrade various lipids. The plant glyoxysome is a peroxisome that also contains the enzymes of the glyoxylate cycle, which is crucial to the conversion of fat into carbohydrate.

The Endoplasmic Reticulum

The endoplasmic reticulum (ER) is a system of membranous cisternae (flattened sacs) extending throughout the cytoplasm. Often it constitutes more than half of the total membrane in the cell. This structure was first noted in the late 19th century, when studies of stained cells indicated the presence of some type of extensive cytoplasmic structure, then termed the gastroplasm. The electron microscope made possible the study of the morphology of this organelle in the 1940s, when it was given its present name. The endoplasmic reticulum can be classified in two functionally distinct forms, the smooth endoplasmic reticulum (SER) and the rough endoplasmic reticulum (RER). The morphological distinction between the two is the presence of protein-synthesizing particles, called ribosomes, attached to the outer surface of the RER.

The Golgi Apparatus

The Golgi complex is the site of the modification, completion, and export of secretory proteins and glycoproteins. This organelle, first described by the Italian cytologist Camillo Golgi in 1898, has a characteristic structure composed of five to eight flattened, disk-shaped, membrane-defined cisternae arranged in a stack. Secretory proteins and glycoproteins, cell membrane proteins and glycoproteins, lysosomal proteins, and some glycolipids all pass through the Golgi structure at some point in their maturation. In plant cells, much of the cell wall material passes through the Golgi as well.

The Golgi apparatus itself is structurally polarized, consisting of a "cis" face near the transitional region of the RER, a medial segment, and a "trans" face near the cell membrane. These faces are biochemically distinct, and the enzymatic content of each segment is markedly different. The cis face membranes are generally thinner than the others. As the secretory proteins move through the Golgi, a number of chemical modifications may transpire.

Important among these is the modification of carbohydrate groups. As many secretory proteins are glycosylated in the ER. In the Golgi, specific enzymes modify the oligosaccharide chains of the glycoproteins by removing certain mannose residues and adding other sugars, such as galactose and sialic acid. These enzymes are known collectively as glycosidases and glycosyltransferases. Some secretory proteins will cease to be transported if their carbohydrate groups are modified incorrectly or not permitted to form.

In some cases the carbohydrate groups are necessary for the stability or activity of the protein or for targeting the molecule for a specific destination. Also within the Golgi or secretory vesicles are proteases that cut many secretory proteins at specific amino acid positions. This often results in activation of the secretory protein, an example being the conversion of inactive proinsulin to active insulin by removing a series of amino acids.

Secretory Vesicles

The release of proteins or other molecules from a secretory vesicle is most often stimulated by a nervous or hormonal signal. For example, a nerve cell impulse triggers the fusion of secretory vesicles to the membrane at the nerve terminal, where the vesicles release neurotransmitters into the synaptic cleft (the gap between nerve endings). The action is one of exocytosis: the vesicle and the cell membrane fuse, allowing the proteins and glycoproteins in the vesicle to be released to the cell exterior.

As secretory vesicles fuse with the cell membrane, the area of the cell membrane increases. Normal size is regained by the reuptake of membrane components through endocytosis. Regions bud in from the cell membrane and then fuse with internal membranes to effect recycling.

Sorting of Products by Chemical Receptors

Not all proteins synthesized on the ER are destined for export. Many, such as the hydrolases in lysosomes, remain inside the cell; others become anchored in the membrane of internal organelles or in the cell membrane. It is presumed that each protein has some type of marker that fits a specific location in the cell. Proteins synthesized on free ribosomes have segments that bind to specific receptors on the outer membrane of mitochondria, chloroplasts, or peroxisomes, allowing these proteins to be taken up only by these organelles.

In the case of proteins synthesized in the RER, both the hydrolases destined for lysosomes and the secretory proteins are found initially in the same portion of the ER lumen. Studies have shown that these can be distinguished on the basis of their carbohydrate residues. The carbohydrate residues of lysosomal enzymes become modified in the cis-Golgi by the addition of certain phosphate groups.

This critical modification allows the enzymes to bind to specific receptors on the membrane of the Golgi, which then directs them into vesicles leading to a lysosome rather than a secretory vesicle. In the lysosomes, proton pumps create an acidic environment that causes the release of the lysosomal enzyme from the membrane-bound receptors. Much of this sorting activity is mediated by coated vesicles containing the same fibrous outer protein, clathrin, used in endocytosis. These sorting vesicles also contain associated smaller proteins.

Bibliography

Aw, S. E.: *Chemical Evolution*, Singapore, University Education Press, 2006.

Brock, H. : *History of Chemistry*, New York: Norton, 2012.

C. Elliott: *Biochemistry and Molecular Biology*, Oxford University Press, Delhi, 2015.

G Singh: *Chemistry of Amino-Acids and Proteins*, Discovery, Delhi, 2017.

G.P. Garg: *Biochemistry of Nucleic Acids*, Discovery Pub, Delhi, 2023.

Geis, I.: *Chemistry, Matter, and the Universe*, Menlo Park, Ca., W.A. Benjamin, 2006.

Haber, L. F. : *The Chemical Industry: 1900-1930, International Growth and Technological Change*, Oxford: Claredon Press, 2001.

Harsh Bhaskar: *Basic Facts on Biochemistry*, Campus Books, Delhi, 2020.

Jagdish Chander and Anil Kumar: *A Comprehensive Text Book of Applied Chemistry*, Abhishek Pub, Delhi, 2019.

Lata Bhattacharya: *Animal Biochemistry*, Discovery, Delhi, 2023.

Maya Majumdar: *Conceptualization of Child Nutrition*, Adhyayan, Delhi, 2016.

Nisha Khalsa: *Essentials of Biochemistry*, Aavishkar Pub, Delhi, 2018.

Patricia Trueman: *Nutritional Biochemistry*, MJP Pub, Delhi, 2017.

Pooja Bhagwan: *A Handbook of Inorganic Chemistry : Nuclear, Atomic, Aqueous, Alkali and Alkaline*, ISPA, Delhi, 2015.

R. Ling: *A Textbook of Dairy Chemistry*, J V Publishing House, Delhi, 2018.

Ram Naresh Mahaling: *Basics of Biochemistry*, Anmol, Delhi, 2018.

Rashmi Deshprabhu: *Child Development and Nutrition Management*, Book Enclave, Delhi, 2014.

Richard Myers: *Basics of Chemistry*, Atlantic, Delhi, 2017.

S S Dogra: *Textbook of Environmental Chemistry and Pollution Control*, Swastik Pub, Delhi, 2017.

S. Banerjee: *Cell and Resource Biology*, Dominant, Delhi, 2014.

S.K. Prasad: *Biochemistry of Carbohydrates*, Discovery Pub, Delhi, 2023.

Shankar Ojha: *Aquaculture and Nutrition and Biochemistry*, Agrotech, Delhi, 2016.

Smriti Ganguly: *Biochemistry of Biomolecules*, Pearl Books, Delhi, 2017.

Stewart Truswell: *Essentials of Human Nutrition*, Oxford University Press, Delhi, 2018.

Sudharmai Devi: *Analytical Procedures in Soil Science and Agricultural Chemistry*, Agrotech, Delhi, 2014.

Sunetra Roday: *Food Science and Nutrition*, Oxford University Press, Delhi, 2017.

V K Chhazllani: *Dairy Chemistry and Animal Nutrition*, Manglam Pub, Delhi, 2018.

Van Antwerpen, F.J. : *The Origins of Chemical Engineering, History of Chemical Engineering*, Washington D.C.: ACS, 2000.

Vasudev Bhatnagar: *Cell Science and Technology*, Campus Books, Delhi, 2019.

William Alec : *The Chemical Industry*, London: Longman Group Limited, 2001.

Index

H

L

M

N

O

P

R

S

T

V

W